THE CAPTAIN

COMMANDER

Leading Fearlessly Above Your Rank

Written by Captain Dan "HH" Hochhalter

Library of Congress Control Number (LCCN): 2024901395

Paperback ISBN: 979-8-9899208-0-8

Cover designed by Jon Alderman

Page break image design: Freepik

Chapters

"Just because you're a Lieutenant,

doesn't mean you need to act like one."

- Captain Dan "HH" Hochhalter

Dedicated to the freedom of expression,

because this book will embolden the strong and offend the weak,

without pandering to either.

Introduction

Fear drives the majority of decisions made in today's litigious environment. Leaders are afraid of losing their jobs over their authentic words and bold actions, so they play it safe and don't truly lead with their heart and gut. This is crippling the United States, and these leaders don't deserve to hold positions of authority. With globalized markets and volatile world order, we need fearless leaders more than ever before, in every sector. This is not a book on military leadership; it is a call for those ready to step-up to the plate of leadership and take bold actions, even if it's before their rank or position gives formal authority to such brazen action.

How I write this book is partial evidence of the way I lead. I will not shy away from stating in simple terms the keys to my success as a leader. If I wrote this book under the cloak of political correctness or cultural sensitivity, I would be spoiling the "secret recipe," and those choosing to implement my suggestions would not find success. Therefore, read with the understanding that what I say and how I say it, is for your benefit. So get over yourselves, your sensitivities, your self-righteousness, and buckle up for some truth in leadership.

If I am going to read a book (most often *listen* while I commute), I want it all to be valuable. I also don't want a lot of unnecessary words or cliché phrases. Ain't nobody got time for that... So in an effort to not waste the reader's time I have decided to format my writing to always fit into three categories that I feel are worth your consumption of the entire book:

1. <u>Leadership Tactics:</u> theories, practices, suggestions, and nuggets of wisdom that have worked for me (e.g., immediately after meeting someone new, write down their name and two unique things about them to help you remember them by name).

2. <u>Gouge:</u> free glances at my homework, willingly sharing "how I did it," "what worked," and handing over blueprints without expecting anything in return (e.g., giving you a peek into how we planned the most audacious air traffic control training event that became a national template for others to follow).

3. <u>War Stories:</u> personal stories about successes and failures (no, they don't have to be from "war/battle"—it's just an expression we use in the military for when we're about to tell a personal story) (e.g., relaying the story about the extreme lengths I went through after my car suddenly broke down, just to make it on time to a military exercise 1,428 miles across the country).

I won't outline which of these three I am giving you at the time, I'm just providing insight into how I categorize my thought process to then organize my writing. This should help you understand why I took the time to write what I wrote.

Tip: Don't give up on reading after the first couple of chapters. This is one leadership book you don't want to fickly discount. I've got to set the stage appropriately, so don't be a damn quitter. I've tried to ensure I don't waste your time in how I write. Additionally, I have no "target audience" with this book. Whether you're a civilian or servicemember, currently in an official leadership position or trying to lead without the title, this book has lessons that will make you and those around you successful in life. So be smart and throughout the book find ways to apply the reading to your personal circumstances. There will be no spoon feeding. Be creative and adjust the principles to fit your current situation, so you can live your own unique story.

INTRODUCTION

Acronyms: Like most servicemembers, I use a lot of acronyms. I will spell each acronym out the first time and sometimes later in the book even if it's a repeat, just for clarity.

Disclaimers: All content is from personal experiences. Most often names are simply left out, but any names I do mention have been changed to pseudonyms, unless it's a direct quote. The views expressed in this publication are those of the author and do not necessarily reflect the official policy or position of the Department of Defense or the U.S. government. The public release clearance of this publication by the Department of Defense does not imply Department of Defense endorsement or factual accuracy of the material.

Chapter 1

How I Got Here

How did I get here? Kneeling over a bed in an empty hotel room in Long Island, New York, crying like a little child. My family of six was technically homeless at the moment. We had sold our house in Idaho, moved into a 700 square foot apartment in Oklahoma City for a summer, and were now living out of a single hotel room in New York. How the *hell* did I get here?

Personal ambition had gotten me there.

I joined the Air Force back in 2007 to become an Air Traffic Controller, finish earning my bachelor's degree, and have more babies for free. The plan was to complete my six-year enlistment having accomplished those things, and then separate from the military and work for the Federal Aviation Administration (FAA) earning the big bucks as a civilian controller. Well, everything had nearly gone exactly as I had planned.

I had just finished #1 in my class at the FAA Academy's Ten-Eleven-Twelve Radar Assessment (TETRA) course. My reward for such a ranking was getting sent to New York TRACON (Terminal Radar Approach Control), or "N90" as the facility is named. N90 houses the highest paid air traffic controllers in the country, due to their locality pay being so high and congress approving N90 to increase their Control

Incentive Pay (CIP) to 1.5 times the amount of any other "hard-to-staff" FAA ATC facility. They control all the radar approaches and departures of LaGuardia, Newark, JFK, and forty-seven other satellite airports. It's insanely busy.

So there I was…with my ATC credentials, master's degree, wife, and now four children—having been hired by the FAA to control from the highest paid ATC facility in the country. Mission accomplished. Every goal I had set back in 2007 had been fulfilled.

Yet, I was living out of a hotel room on Long Island, with everything I owned in a U-Haul parked outside; my wife and children out to lunch, as I hit my emotional wall…alone in that hotel room. None of it felt right. It just felt wrong to be in New York.

I was sobbing, because for the first time in my adult life I didn't have an answer for the way forward. Taking a family of six to Long Island with no prior housing plans, with an income that wasn't going to be able to afford anything on a single-income household until I earned my ratings (average of 2.5 years), was a reality we never anticipated. I even remember the waitress at a diner on Long Island the day we arrived asking where we lived. I told her we were living out of a hotel room and had a week to find housing, to which she replied in the thickest New York accent you can imagine, "What…are stupid or somethin'?"

Apparently, yes.

We had tried everything in an attempt to find any type of housing. One-bedroom apartments, rundown rentals, putting in offers on the cheapest homes (still well outside our financial reach). We even met with an older lady to negotiate living in her basement temporarily. None of it was falling into place, and after seven days of defeat after defeat, I finally broke down. I didn't have a place to put my family. We had previously moved all over the country and been able to find housing easily each time, but Long Island was another planet. I have never felt so worried as a husband and father, not to mention feelings of failing my family.

The instructors at the FAA Academy had warned me.

As I was nearing the end of my TETRA training at the academy, many of the instructors had gotten to know me well (it was a total of 4.5 months of arduous training). They knew I had a family of six and was a single-income household at the time. Knowing the majority of the class was going to N90, they kept warning me of the difficulties ahead. Not just financially, but also what to expect at work. There are over two hundred controllers at N90. Once I arrived, I would be at the very bottom of the seniority totem pole, which means I would have the worst days off and would work every holiday…for at least the next ten years. I was told to say goodbye to seeing the looks on my children's faces come Christmas morning, because I'd be staring at a radar scope instead. By retirement, I'd likely be on my third marriage, have four vices, look like I'm eighty when I'm only fifty-six, and be in debt due to the vices and ex-wives.

Is this what I signed up for?

The final assessment to pass TETRA involves three intense control simulations over the course of two days. They bring in ATC facility managers from the busiest facilities to evaluate you. If you pass, you go to a 10-, 11-, or 12-level facility (12-level is the busiest and highest paid). If you fail, they send you to a 7-, 8-, or 9-level facility (less busy, less pay) for further training and assessment.

The day before this final evaluation one of the instructors pulled me aside on a break. You could tell he was truly concerned about me taking my family to N90. He explained the immense stress that type of facility would put on me as a controller, as well as the hardships my family would experience living on Long Island. At the end of his concerned speech, he leaned in and whispered, "You should fail tomorrow's assessment on purpose. That way you're guaranteed not to go to N90 and starting out at a sub-10 facility isn't bad."

Furthermore, that same day I happened to run into the faculty member who is over facility placement, and in confidence he showed me the list of facilities currently available if I failed and had to pick from a sub-10 list. Most of them were pretty good facilities in great locations.

I had never in my entire life contemplated failing something on purpose. I'm ashamed to say, I gave it the rest of the business day to think it over.

On my drive home, I pulled over to the side of the road and shut off the car. What was going on?! Fail on purpose?! Up to this point in my life I had always been a top performer, and prided myself on never giving up, even when every odd pointed toward failure. Was I going to compromise my personal integrity in order to save my family and myself from the pitfalls of N90? Wow...this made my head spin. I couldn't believe what was happening.

Then I remembered what had gotten me to this point in my life and career: giving it my very best, every day, every damn time. That was not about to change, come what may.

I got home and told my wife about the conversation with the Instructor and the interaction with the facility placement manager. She agreed completely, failing on purpose wasn't an option. She was all in.

Over the next two days of evaluation I ended up setting an FAA Academy record by scoring perfect one hundreds on all three simulations. I was told that had never happened before.

•◆————————————◆•

Just one week later, I am sobbing on my knees in a hotel room on Long Island, wondering how I was going to provide for my family. Then I received a text that would forever change the trajectory of my life.

It came from a friend I had made while attending Undergraduate Air

Battle Manager Training at Tyndall, AFB (Air Force Base) Florida. He had since then taken a job at the Air Force's Officer Training School (OTS) and was currently serving there as the Director of Operations (#2 in command). I had stayed in touch with him over the years and when I heard he had taken an OTS assignment, told him I had always wanted to return to OTS to instruct (by this time in my career I had commissioned as an officer, but you'll read about this later. You'll also read and understand that at this point I had taken a break from the full-time Air Force to pursue my FAA dreams but was still in a reserve status with the Idaho Air National Guard.)

His text said that OTS was hurting for Instructors, and he was wondering if I wanted to take two to four years of active orders to come instruct down at Maxwell AFB, Alabama.

This text came on the hardest day of my life, during the hardest moment as a husband and father.

I replied asking if he was serious and if so, how soon could I start. He confirmed he was serious and said I could start next week. I collected myself and then went to meet my wife and kids at the place they went for lunch. I explained the text I had received and the opportunity down in Alabama.

[Note: It is painful sharing such personal things with the world, but I do this hoping my story will resonate with someone and possibly help them, the way I have been helped by others. Sharing my wife's emotions and thoughts is deeply personal, but necessary for this part.]

By now, my wife had gotten pretty beat-up emotionally. I pulled her away from her dream home, in her dream location in Idaho, from her deep-rooted friends, all to pursue *my* dream of becoming a highly paid FAA air traffic controller. Then I moved her and our six-member family into a tiny apartment in OKC, made her endure the hot summer with four kids and nothing to do, and then showed up to New York without a place to live. She was emotionally done. She couldn't handle anymore, and this is a woman who gave birth without family around during the middle of my seven-month deployment to Iraq, and never complained.

She is as tough as they come, but this had broken her, and she said, "Dan, this one's on you. I will follow you wherever you take us, but I'm not making this decision with you."

For some reason the thought of instructing at OTS removed the entire world of weight that was resting on my shoulders. It just felt right in my heart and in my gut. I made the decision to jump in the U-Haul the very next day and start driving us south toward Alabama, leaving my FAA dreams in the rearview mirror.

Fast forward to me instructing at OTS. My family is settled in the beautiful big home we bought in Alabama, the kids are loving school and making new friends, my wife is involved in the community, and I am crushing it as an OTS Instructor. It ended up being the right choice, one hundred percent. Nevertheless, the jolt of that journey took a toll on our family and my marriage for many years to come. I must move on to progress the point of this part of the book but know there are many scars from that experience.

———◆———

One day while working at OTS I approached a trainee wearing an ATC badge on his uniform. This means he was prior-enlisted and was selected to commission as an officer (which is the route I took). I gave him a hard time about something (can't remember exactly what it was), and then commented on his ATC badge. I said some smart remark and then walked away. These types of interactions were common throughout the day at OTS; finding things to add stress to the trainees' daily lives.

Toward the end of OTS, the trainees are given more privileges, to include being allowed to work out in the OTS gym where most of the Instructors exercised. While I was working out one day, this same trainee approached me and asked if we could talk. He ends up explaining his pedigree in ATC and that he recently transferred from active duty to the full-time Air National Guard in Wyoming. He also mentioned that they were going to be looking for a new commander soon, and wondered if I were interested.

I literally let out such a loud laugh in the gym that it caused everyone to look over.

I had *just* pinned on Captain, and squadron command was reserved for Lieutenant Colonels, which was two ranks above Captain. What was this guy smoking?

He said, "You can't win, if you don't play…sir."

I brushed it off, and also told him how much I was enjoying Instructor duty, but that we could exchange numbers for networking purposes since the ATC world was a small one.

Over the course of the next year he kept in touch with me. He even returned to Alabama to attend a fellow Wyoming guardsman's graduation and asked to go out to breakfast. He talked about how his ATC squadron was in a bit of disrepair and the current commander was under investigation. He predicted the commander would be removed. He kept alluding that they might consider hiring a captain, given my background in ATC and Air Battle Management. Once again, I brushed him off. That just never happens. You never see a Captain Commander.

When the job announcement finally posted, he texted me again encouraging me to apply. Keep in mind, I was only a year-and-a-half post the New York crisis. I had promised my wife up to four years of stability in one location (we had moved a lot up to this point). I was loving my job at OTS and my kids were young enough that they seemed to have come out on the other side of that ordeal, unscathed. My wife would kill me if I even mentioned this!

So I went home and mentioned it to my wife…

At first, she was hesitant. The advertisement was open for a couple of months, so we didn't have to make any immediate decisions. Then I received a phone call from the Operations Group Commander in

Wyoming (the boss over the ATC Commander). He said he had heard about my resume and was interested to know more.

My wife agreed to at least fly to Wyoming to check out the unit and area before I applied. That trip went really well, and we both felt good about at least throwing our hat in the ring.

I applied. I interviewed. I got the job.

I couldn't believe it. Yes, my resume was very strong and perfectly diverse for this position. Yes, I crushed the interview board and walked away feeling like I couldn't have done any better. Yes, I was excited about the possibility of returning to my enlisted roots in ATC now as a commander and leading much sooner at that level than I had anticipated. However, I truly didn't think they would offer me the position. Nevertheless, unknown to me, my future bosses saw potential in me that I didn't even know I had, and they would end up helping mold me into the leader that organization needed.

Three months later I was ceremoniously handed the squadron guidon of the 243d Air Traffic Control Squadron (243d ATCS).

And *that*…is how I got here.

This is the story of how I took the 243d ATCS from the bottom of the pile, to the number one ATCS in the nation. My story can double as a leadership handbook for anyone in my same unique situation of being appointed "boss" without the commensurate rank, or for those that *want* to lead above their current "rank" and position that might not possess the official authorities. In either case the stories, antidotes, and theories I express in this book work for propelling individuals and organizations toward excellence. They worked for us, because of the people of the 243d ATCS who took a chance on a Captain Commander and ran next to my side as we charged hill after hill, conquering each new challenge and leaving no doubt concerning our purpose and promise. I offer my experiences to you and hope for your success as a leader.

Chapter 2

Why You Give a Shit

I never understood how someone could teach a subject they had no real-world experience in. I'm not talking academic experience. I'm talking actual "been there, done that" kind of experience. Preferably, you've also had an array of results surrounding those experiences; not one dimensional. So I want to start out by providing you the reason why you should give a shit about what I have to say. Also, I need the reader to understand you don't have to be an officer in the military or hold an official leadership position in order to apply these principles and lead right where you currently stand.

I am writing this book as a thirty-seven-year-old. I am far from the seasoned leader that typically writes this type of book. My intent behind writing at this age and time in my career is to provide a practical path toward being a damn good leader...now! Not when you're in your fifties, not once you've obtained that big promotion, not after you've finished that next level of training. Look around at the rise of younger leaders in nearly every profession: tech, sports, social media, marketing, politics, business, etc. More than ever before in history we are seeing younger leaders, and as a young leader in a position usually held by someone older and at least two ranks higher than me, I think I have valuable information and experience to share on this matter. I've "punched above my weight," so to speak…and now I want to help others learn how to "lead above their rank." I boldly dare to suggest that youth and wisdom are not mutually exclusive.

This is not a book on "how to lead in the military." The principles I share are universal, and if you can't use your imagination and figure out how my military experiences apply to your civilian leadership/career, then you have discovered an important truth about yourself: you're not meant to be a leader. Not that this is a bad thing. In fact, I really hope you don't apply the "little engine that could principle" by "thinking you can" and end up ruining organizations and peoples' careers practicing being a shitty leader. You just need to find your passion and leverage your skills in other ways (see "Chapter 16: Your Authentic Recipe").

I am taking substantial risk to my career by publishing a book before I am cloaked by the safety net of a retirement pension. You'll come to find out by reading this book that playing it safe isn't my style, nor do I believe it should be any leader's style. My hope is that this book will inspire all who read it to be bold and make the necessary changes in themselves and their organizations to better take care of the people that make the magic happen on a daily basis. To wait until I was retired to write this, would go against everything I am trying to encourage in this book.

Will there be a time, years down the road, where I look back at the things I wrote and realize I was slightly naïve? I sure hope so, otherwise I will not have grown. However, there is something special about young, passionate, driven leadership that every organization absolutely needs in order to be successful long-term. So I can appreciate my own current naivety, because it's what's helped propel my team, and I passed those who "know better" and are therefore unwilling to take the necessary risks.

If you are "well-seasoned" (old) and reading this book, then maybe you could learn some new tricks to help keep you relevant and maybe rekindle that fire you used to have as a younger, more ambitious you.

I write mainly from my experiences as a commander, with the

authorities associated. I understand that leading above your current rank also means leading without the full authorities you may need in order to actuate change in your organization or effectively pushback on counterparts. I can empathize with your frustrations; having had the right "recipe" in prior positions, but no "kitchen" in which to cook. Stick with me throughout this book and try to use the experiences I've had *with* command authority to still help lead others around you to a better version of themselves, and thus increase the strength of your organization overall. I didn't just start applying these principles the day I became "the boss" …which is part of the reason I became "the boss" at such an early rank in the military.

I will write as if my way is hard fact, but if you're reading this book, you've likely read other books on similar topics and should know by now that leadership isn't a science, it's an art. So this is probably the only time I'll admit that my way isn't the only way. It's just the *best* way (see "Chapter 8: Humility is Overrated").

Before I start wanting to believe in what someone is saying, I want to know their background and credentials. So, we'll start there. Some might consider this next part bragging; I consider it a job interview for earning your time to read my book.

I am currently serving as the Commander of the 243d Air Traffic Control Squadron out of Cheyenne, Wyoming. I am a Captain in the United States Air Force. I joined in 2007 and was "Active Duty" (more appropriate term is RegAF: Regular Air Force) for seven years as an enlisted Air Traffic Controller (ATC – 1C1). My first duty station was Luke AFB, Arizona, where I grew as a controller. In 2010, I was deployed for seven months to Joint Base Balad, Iraq, where I controlled the busiest airspace during the war in the Middle East. Shortly after returning to Luke AFB, I was ordered to the ATC schoolhouse in Mississippi to be an ATC Instructor. After a couple years of teaching at the schoolhouse, I decided that as a Staff Sergeant (E-5) with five dependents and a Master's degree, that I could do better. So I applied for a position in the Idaho Air National Guard (ANG) as an Air Battle Manager. I got the job and transferred seamlessly from RegAF into the full-time ANG as dual-status air technician (which essentially means I get paid by a different pot of money, can be ordered by both the President of the United States and Governor of the state, and am still a full-time Airman wearing the uniform every day and working on base).

The position I had been hired into required me to be a commissioned officer. So I was sent to Air Force Officer Training School (OTS) at Maxwell AFB in Alabama where I earned my commission as a Second Lieutenant, with nine-month follow-on training at Tyndall AFB in Florida to become an Air Battle Manager (ABM – 13B). I worked as a full-time ANG ABM at Mountain Home AFB in Idaho for four years before jumping on the FAA Academy opportunity you read about earlier, followed by taking the assignment back at Maxwell AFB to be an OTS Instructor.

While I was at OTS as an Instructor, I stumbled upon that ATCS Director of Operations who got word back to Cheyenne, Wyoming, about me. As you know, I applied to the Air Traffic Control Squadron (ATCS) Commander (CC) position and was hired, because they couldn't find anyone else… I also had a bulletproof resume, crushed the hell outta the interview, and gosh darn it…people like me. So that's how I ended up in my current position. There are lots of gaps in that nutshell that I will fill in along the way.

◆——————————————◆

The title of this book is contradictory if you understand how the Air Force rank and position structure works. You just don't see captains commanding a squadron. So running across a Captain Commander is like finding rocking horse shit. I'm not saying I'm the first or last Captain Commander, but it is extremely rare (have I mentioned "Chapter 8: Humility is Overrated" yet?).

As much as I'd love to write this book as if I were speaking to a seasoned Air Force veteran by using jargon and acronyms all over the place (this would be much easier since it takes less time and is more effective), I am hoping to influence more than just 507,876 people (number of those currently serving in the Air Force). So I will take moments to explain "Air Force-isms" throughout the book.

Let me start with Captain. Officers in all of the U.S. military branches are ranked from O-1 (Second Lieutenant; aka: "Butter bar") through O-10 (General; aka: "Four Star General"). Enlisted are similarly ranked E-1 (Airman Basic; aka: "Fuzzy") through E-9 (Chief Master Sergeant; aka: "Chief"). I am an O-3. I can't say "Captain" and it mean the same thing in the Navy, because a Captain in the Navy is an O-6 (the Navy is pretty unique/confusing). It can be perplexing how each branch calls ranks a little different (especially the enlisted), but we all can speak the same language and understand "Os and Es and numbers." We all use the same "E" and "O" pay charts. This is why we call them "pay grades."

So Captain in the Air Force is no big deal. You breathe for four years after being commissioned as a Second Lieutenant and don't do anything stupid...you'll become a Captain (with First Lieutenant at the two-year mark in-between...but a First Lieutenant is just as useless as a Second Lieutenant, so it really doesn't matter).

There is something called an O-3E. This means an officer served at least four years as an enlisted servicemember before losing what little sense they did have by commissioning as an officer. This is me. I was an E-5 (Staff Sergeant) before I decided to give up on logic and become an entitled complicated officer.

So that's rank. Now onto positions. Commander is a position at each major tier in the Air Force. Although the smallest unit is a Flight, they don't make up a major tier/unit by themselves. The officers leading Flights don't have any "G-series orders authority" (not to go down this rabbit hole too far, but pretty much "G-series" gives Commanders the authority to create intent and policy, hire/fire, promote/demote, and administer non-judicial punishment up to an Article 15 under the Uniformed Code of Military Justice (UCMJ)). So the first unit level where G-series Commanders reside is the Squadron (typically between sixty to two hundred Airmen). Then from there it goes Group (about four Squadrons) and then Wing (about four Groups, totaling sixteen Squadrons). There are more layers above Wing, but that's as far as I'll explain for now. Just understand that most Air Force bases (not all) have one Wing, so that Wing Commander is essentially the "Base Commander." On my current base in Cheyenne we have one Wing, four Groups, and thirteen Squadrons. Squadrons can vary in size, but all Squadrons on my base are just under one hundred Airmen strong.

In the Air Force, Squadron Commanders are typically O-5s. If you're a civilian and that last sentence made sense, then you've passed the test! For the other branches, an Air Force Squadron Commander is equivalent to an Army or Marine Corps Battalion Commander, and again the Navy is complicated, but it's the equivalent unit size that the actual rank of a Navy "Commander" O-5 would be commanding (I guess that kind of makes sense now that I write it out…maybe I've been wrong about you all along, Navy…).

So now you understand my current *rank* (Captain/O-3E) and *position* (Squadron Commander), and the meaning behind the title of this book: *The Captain Commander: Leading Fearlessly Above Your Rank.*

The reason behind the title and the content is to provide a unique perspective on what it's like getting that crazy chance to actually lead an organization before your rank typically allows, and how the leadership principles I used *before* getting this position which I still applied *after* assuming command, contributed to our success. This means you can start *now* to lead where you stand! The quote at the beginning: "Just because you're a Lieutenant, doesn't mean you have to act like one," I coined while an Instructor at Officer Training School. I will dive into this quote more later, but it's pretty simple; you need to be performing and acting at least one rank above your current rank, or else why the hell would we promote you and give you influence over more people?

The following are highlights of my resume. I am not going to drill down on these and provide a lot of explanation. You'll find I actually do believe humility is a needed quality of a leader, so don't take this as me putting myself above anyone else. I just appreciate knowing where my information is coming from, and since I won't use any other sources in this book besides my own experiences and a few cited quotes, I think I owe it to the reader to allow you to judge whether or not I am a valid source on any of the topics.

My credentials:

- Master of Science in Leadership
- Bachelor of Science in Business Administration
- Three Associate degrees (General Studies, ATC, and Instructing)
- Basic Military Training – Honor Graduate
- ATC Tech School – Top Graduate
- Airman Leadership School – Levitow Award
- Officer Training School – Distinguished Graduate
- Undergraduate Air Battle Manager Training – Yukla Award (missed Top Graduate by .02% to my best friend in the class...he earned it and it ain't easy beating me; great job, Mike)
- Air Traffic Controller of the Year – Luke AFB
- Company Grade Officer of the Year – Idaho ANG
- OTS Instructor of the Year
- SrA (E-4) Below-the-Zone
- Staff Sergeant (E-5) first attempt
- Seven-year enlisted Air Traffic Controller (Luke AFB, AZ; Joint Base Balad, Iraq; Keesler AFB, MS)
- Iraq War Veteran (Operations IRAQI FREEDOM and NEW DAWN)
- Master Instructor - Air Education and Training Command
- Master Air Traffic Controller
- Senior Air Battle Manager
- Advanced Air Advisor
- SWAT Basic Tactical Operator
- 2021 Unit of the Year (243d ATCS)
- 2021 & 2022 Cheyenne Trophy (243d ATCS)

- Voted by the ATCS Weapon System Council as #1 of 10 ANG ATCSs to convert to a Combat Airfield Operations Squadron (CAOS) and one of only two units selected to stand up a Landing Zone schoolhouse

*Strats (ranking among peers given by commanders): I will not share specific strats, since this information could hinder relationships. However, I have always received the ranking that I would expect out of my own leaders.

My Captain Years (four years, seven months):

- OTS Instructor (eighteen months; a few months as 1st Lt)
- Squadron Commander (three years)
- Air Advisor deployment (six months)

*(As of the publication of this book, I had promoted to Major and therefore will not include anything past the summer of 2023 in my experiences I share, so that it's a true "Captain's perspective.")

In short, I have sixteen years of full-time service in the Air Force (as of 2023), a master's degree in leadership, have taught leadership at multiple Air Force schools (attained "Master Instructor" credentialing), have been in the top five percent of every measurable thing I've done in the military, was hired to lead an Air Force squadron two ranks above my pay grade, and have taken that squadron from the bottom of the pile to the number one position. I have a passion for leading people and teaching, and I boldly offer you my experiences and advice, but I humbly still have so much to learn.

◆———————————◆

Who are the 243d Air Traffic Control Squadron (243d ATCS) "Red-tailed Hawks"?

We are a seventy-two-hour mobile air traffic control and airfield weapon system. This means, within seventy-two-hours of being ordered to deploy, the entire unit (personnel and equipment) can be recalled and flying to anywhere in the world to set up air traffic services. Whether it be to Florida with our mobile tower and controllers to temporarily supplement an airport taken out by a hurricane or setting up and controlling an airfield on some remote island in the Pacific in order to refuel fighters and resupply cargo planes on dirt strips during the next war. The 243d ATCS is equipped to open and control an airport no matter the scale, no matter the surface, no matter the location. We are one of eleven such units in the entire U.S. Air Force; ten ANG and one RegAF.

What have I done that's so special?

Like I mentioned before, there are plenty of other leaders with much stronger resumes, higher degrees, better war stories, and years of experience that outweigh my accomplishments within the sixteen years I have served thus far. However, this book is about leading above your current rank, so I will primarily focus on my first three years as a Captain Commander.

[Note: At exactly my three-year mark of command, I deployed for six months as an Air Advisor on a Mobile Training Team to help stand up the new Qatari Air Operations Center. Upon returning from that deployment, I promoted to Major [O-4] and have continued serving as Commander of the 243d ATCS as of the publication of this book.]

One of the benefits of the ANG versus RegAF is that I can stay in command longer than the typical two to three years of those commanding in the RegAF. Even as a Major, I am still one rank below what I should be as a Squadron Commander, so I have many years of eligibility left in the seat. We'll see how long it takes me to get fired.

I could share many stories and profound experiences from my previous twelve years in the Air Force before assuming command, but again I want to concentrate on the principles of leadership that I believe worked to turn the 243d ATCS into the top performing team that it is. Yes, I've deployed, am a war veteran, and have been attacked by mortars with close calls from those landing inside the wire, aka: Fobbit attacks (no "outside the wire experience"). Yes, I've missed crucial life moments including the birth of my child due to my military obligation. Yes, I've had heartbreaking moments of losing loved ones.

No, I am not a combat veteran. No, I haven't directly killed anyone. No, I'm not pretending to be a Navy SEAL. This is leadership, and much like art and music, you have to find the artists and musicians that speak to your soul and help you find passion in life. My style might not be the one that does that for you, but I thank you for giving it a try.

The culmination of three years of commanding, leading, molding, mentoring, and taking risk came down to the execution and the post-exercise success wave of "Triple ACE." Triple ACE was a homegrown proof-of-concept training event that my squadron created from a very ambitious claim, which turned it into a success that is now being recognized by top Generals and emulated by the Air Force ATC/Airfield Operations community.

ACE stands for "Agile Combat Employment" and is the new frontier of warfare; meant to be more light, nimble, and lethal. It's a drastic evolution from how we fought in the Middle East, and it's in preparation for a potential future conflict in the Pacific. It's as much a mentality as it is a practice, and Triple ACE was our interpretation of what our Generals were preaching concerning ACE and MCA (Multi Capable Airmen).

The next couple of paragraphs might make your eyes go crossed with numbers and acronyms, so push through, but in short, during Triple ACE the 243d ATCS singlehandedly controlled 1,041 air operations between six locations simultaneously every day, over the course of three days. This more than tripled the previous record set by an ATCS of 316 operations over four days. "Operations" are take-offs, landings, low approaches, etc. Moreover, a C-130 offloaded an empty R-11 Fuel Truck from its cargo bay (Engines Running Offload – ERO), then took fuel from that same C-130 and dumped it into the R-11 Fuel Truck while engines were still running (Wet Wing Defuel), and lastly that R-11 was then used to HPR (Hot Pit Refuel) multiple UH-60 Blackhawks (refuel the helicopters while their engines were also still running). That entire ACE process shaved ninety minutes off aircraft downtime. This also marked the first ever HPR in Wyoming ANG history.

We headquartered all of this out of one location (MOB – Main Operating Base) from which we started and ended each day, while expanding to six locations every day (FOBs – Forward Operating Base), for three days straight. It's called "hub-and-spoke" (or "MOB-and-FOB") airfield operations, which entailed us deploying small units from the "hub" (MOB) to the many "spokes" (FOBs) scattered across Southeast Wyoming. We would infiltrate (infil) and exfiltrate (exfil) the landing zones via a variety of means at the different locations (helicopters, trucks, by foot, etc.). Rapidly setting up the airfields by surveying and marking the landing zone, we would then control the chaos of demanding air operations for a few hours, before swiftly returning to the "hub" (MOB).

We also proved the concept of a new nationally recognized "tactical wheel" approach system, had controllers *immediately* control at airports they had never operated before (versus the standard deployed spin up time of seven to ten days), and provided our own security with a homegrown team of specialized operators from within the squadron.

The leadership that led to the success of Triple ACE will be explained in more detail throughout the book, since most of it doesn't make sense to the reader yet. It all has a deeper purpose of demonstrating the leadership principles that allowed the above execution to occur, with such a high level of success. So if I lost you just now, don't worry, it will make sense later.

All you need to take away from the past few paragraphs is that the 243d Red-tailed Hawks blazed an unprecedented trail in the world of combat airfield operations. So much so, that we had units baffled as to how we pulled it off, and the most common question after they watched the culminating video was: "How did you do that?" I'm going to explain the answers to that question in this book, and it has nothing to do with military tactics or air traffic control prowess. It's all about leadership.

Please take a moment to watch the video by following the QR code below (one for YouTube and another for a site that works on government computers). Or just search "243 atcs triple ace" on your browser and you'll find it on YouTube that way as well. It's open source and we hope everyone sees it…especially our enemies:

YouTube   Government CPU

The reason I am so proud of what we accomplished with Triple ACE was not the air operations record, groundbreaking tactics, or even the breadth of coverage. It was watching a very ambitious claim I made to my people (that as a single squadron we could control six airfields simultaneously) turn into a historic proof of concept, all because the Hawks took empowerment and an attitude of no failure and turned it into unprecedented execution.

The men and women of the 243d are living proof of the leadership concepts I will outline in this book. Their success is their own, and the only credit I will take is having empowered them to tap into the potential they already had inside, while providing them permission to fail, all while knowing I had their backs. They did the rest and demonstrated gritty resolve.

You should give a shit about what I'm about to say, because I will outline tools that have the capability to drastically transform people and organizations into the fearless leaders and corporations our country needs. It will then be up to you to utilize these tools to be the leader *your people* desperately need. Knowing I helped someone reach a higher potential than they themselves even knew possible, is *the* ultimate reward of leadership, and that is my hope for those reading. Let's reach higher potential and lead above our rank; we owe it to those we are privileged to lead.

Chapter 3

It's Your Fault

Every generation shakes their head at the younger generation in dismay at their behavior and inability to measure up to their "old school" generations' expectations. As a father of four (three of which are teenagers), working as an assistant football coach for the high school varsity team, and my wife as the registrar at the same high school, our dinner table conversations are likely identical to that of my parents'; riddled with over exaggerations of what we saw that day, coupled with a foggy memory of our own teenage stupidity.

However, today's teenagers really do suck way worse than ever before. Like, way worse.

My confidence as a commander and the success of my people, is second only to my confidence that my wife and I are damn good parents and have solid kids. I know that sounds extremely arrogant, but we get more "How did you raise such good kids?" questions from other parents than I get "How did you do it?" questions about Triple ACE, and I'm not writing a book on parenting...but maybe I should. If I did, it would be titled *Parents: It's Your Fault* and the quote at the beginning would be "You're either coaching it or allowing it." That may be another book for a later date, but the short answer to "How did you raise such good kids?" is: my wife and I don't suck as *leaders* of our family and home.

So bear with me as I use a parenting analogy to help drive home this point.

As much as I want to blame the actual teenagers for their lack of respect, personal accountability, and hard work ethic, I have to blame the parents. Parents, how about you take some damn ownership of being a parent, show that you are the adult in the equation, set some guidelines, and be consistent in your follow through? I'm talking about when they're super little, because by the time they're teenagers, applying these principles will be much more difficult if you haven't taught them the basics of respect, discipline, and order (but it's not impossible to instill this when they're teenagers…we do it with eighteen-year-olds entering the military every year). So there is still hope.

I'm not talking about being authoritarian or beating your kids. I know I'm making it sound like we run a boot camp at our house. What I mean is, when you say, "If you don't pick up your toys you don't get dessert,"…then…*you don't give them dessert when they fail to pick up the toys!* Furthermore, you go grab their hand in yours and physically help them pick up every last toy. Every time. No exceptions. Then still *follow through* with not giving them dessert. Never an idle future consequence. Never taking a break because it would be easier to just pick up the toys yourself.

I have, no joke, turned around after having driven down the road in the morning that garbage cans needed to be put out by my boys (their responsibility that I gave them and explained how to do and when), and went down to their room, woke them up, and watched as they moved the garbage cans to the road. I'm not splashing water on them to wake them up or yelling at them for forgetting. I simply throw on the lights, explain that the garbage cans weren't put out, and that they need to take care of it. For anyone reading this that thinks any of that is child abuse, you have abdicated your authority as a parent and fallen victim to societal pressures. It's not abuse to hold your kids accountable. It's called teaching, mentoring, and loving your children. So parents, if your kids suck, it's your fault. These exact same principles apply to leadership.

If you are in any type of official leadership position, it's your fault if things suck or are awesome. It could be as simple as a shift leader at a fast-food restaurant. If service and food are horrible during your shift, it's your fault! If you're a school principal and your students hate the school and teachers, it's your fault! If you're a head coach and your players keep showing up late to practice in the middle of the season, it's your fault!

Leadership is about driving the culture and expectations of your organization. Own it!

•◆————————————◆•

When I was an OTS Instructor, the first morning we received our new set of Officer Trainees (OTs) was an important moment. Every Instructor had their own way of setting the tone on this first day, but I was definitely more on the side of utilizing elevated volume (yelling or "melting faces"). This sounds tough and like the movies, but none of it was effective if the next day I didn't follow through with my expectations and associated consequences...and then every day after that for the next nine weeks of their training. In fact, the first two weeks are the hardest for both the OT and Instructor.

As Instructors, we have to wake up at 0300 (3:00 am) in order to get to the dorms to "kick down doors" (really just pound on their doors) and wake them up at 0430. We are then with them until after dinner. So we are pulling sixteen-hour days the first couple of weeks. However, on the weekends the Instructors don't come in. That very first weekend is a huge relief for us to sleep in and spend time with our families. The OTs, however, are still expected to wake up at 0430 and take accountability and work. I stressed to my flight (in this case a "flight" was a sixteen-member unit of trainees, not an airplane ride) the importance of doing the right thing even when Instructors weren't around. So to ensure my flight was squared away from the very beginning, instead of sleeping in that first Saturday after waking up at 0300 and pulling sixteen-hour days all week, I would wake up at 0300 and show up to the dorms. No other Instructors did this.

I would bring a lawn chair with me and just park it at the end of the hallway with my arms folded, waiting for 0430 Reveille (wake up bugle call) to sound off and watch to see if my flight woke up and stood at the position of attention outside their doors for accountability like they were supposed to. When Reveille sounded, without fail my entire flight would snap out of their rooms and stand at attention while the Flight Leader took roll. The trainees would hazily see me out of the corner of their eyes sitting in my lawn chair, arms folded, and would stand a little straighter at attention while roll was taken. I never said one word. I would just nod my head in approval, fold up my chair, walk back down the hall, down the stairs, and to my car. This let me know I could trust them with more. I would see other instructors' flights not get out of bed on the weekends while I was there to watch mine, and those flights struggled all nine weeks.

However, my coaching and expectations didn't end after that first week. Although at times I was tired and my body and mind would rather not be consistent with the follow through, I knew that if I showed vulnerability to the standards I set, the discipline of my flight would deteriorate. So when the flight was two minutes late marching to their time to eat at the dining facility, they lost their privilege to consume caffeinated drinks until they met every scheduled event on time for an entire week. Conversely, if the class test average was above a ninety percent, I would reward them with off-base privileges that weekend. Over the course of the nine weeks I had with them, I slowly evolved from directive leadership, to coaching, then mentoring, and the last week I was their equal.

I'm not saying to yell at your people or give and take privileges like they are children. The "carrot and stick" leadership tactic has been proven ineffective for long-term intrinsic motivation. Military accession training ("boot camp") is specifically designed with intense yelling and sometimes impossible expectations, just to forge trainees in the fire of stress so that they are prepared for the pressures of their very life-and-death jobs. However, the principles are the same for any leader in any organization: you drive the expectations, and it's your fault if those expectations aren't being met. Utilize the tools you have available to you in order to reward good behavior and correct bad behavior. Every time. Without compromise. Always out of love.

I went straight from OTS Instructor to Squadron Commander. So you would think I might just apply the same outline I used at OTS in order to generate high performing teams. However, I realized I was going to need to adjust my style for my new position of authority.

◆———————◆

Authority. This word means different things to different people, in different contexts. I know not everyone reading this book is currently in a "position of authority" (and maybe ever will). Dr Stefan Eisen put it well, "In today's Department of Defense environment, your span of authority is often less than your span of responsibility. In short, you are charged with mission success while working with people you have no direct authority over." Nevertheless, I believe non-positional leaders (those who don't hold the title, but still have the influence) can apply these same principles. If, at a minimum, they hold themselves to these standards, those around them will notice, and they will have an indirect positive impact on the group.

Regarding G-series Commanders in the military, there is an enormous amount of authority packed into these positions. As briefly and very simply explained in the last chapter, "G-series" means that the commander has the authority to create intent and policy, hire and fire, promote and demote, and administer non-judicial punishment up to an Article 15 under the Uniformed Code of Military Justice (UCMJ). I would like to highlight "create intent and policy." I truly believe the majority of Commanders underutilize this humbling versatile authority.

It's an applause to the decentralized structure of the military. That's right, I said the military is a decentralized command structure. How Commanders properly *leverage* that decentralization is another thing, but it's built to promote a flat organizational chart from which there are few tiers between the lowest ranking Airman of a unit, and their Commander.

In which other countries' military can someone literally dozens of layers down from the four-star General of the entire Air Force (that's me as a Squadron Commander in relation to the Chief of Staff of the Air Force), make decisions such as how to evolve the training of their people to fit the next war? Or the authority to change the way the government's money is spent to better equip their people with what they are asking for? Or realign their organization's chain of command to create a flatter org chart so that ideas from even the lowest ranking member have an expedited route to the top?

These are just small examples of things I have done as a G-series Commander that have revolutionized how the 243d ATCS conducts business. I didn't ask for permission. I didn't wait around for instructions on how to build lighter ATC equipment packages that could shrink our airfield setup time by eighty percent. I didn't ask my people to come up with innovative ideas on how to be more ACE (Agile Combat Employment, aka: "light and nimble") and then only listen to my senior advisors. I wasn't going to wait two years for the Air Force to approve a new squadron patch depicting our newly voted mascot before I wrote a policy letter approving its wear on Fridays and during exercises. For all of these things I leveraged my authority, often with a lump of risk and *maybe* a half teaspoon of "give a shit," and made it happen.

If you want to be a leader, then lead, don't manage.

Yes, "It's Your Fault," but that means if you have an amazing organization that is extremely successful, that's your doing as well. It takes owning your job as the leader and empowering your people to be revolutionary.

My last example for this chapter involves a painting I found just a few months into my command at an antique shop in Fort Collins, Colorado, (I live in Fort Collins and drive forty-five minutes up to Cheyenne every day for work...I'm just not tough enough to live in Wyoming). It was my wife who actually stumbled upon it while we were out as a family window shopping. It immediately resonated with me, and I knew exactly where I would put it.

It is a painting by Robert Schoeller, simply titled "George Washington." It portrays Washington directly facing the viewer, holding a quill in his hand which is also extended outward toward the viewer, and The Constitution laying on a wooden table next to him with an inkwell. His eyes look directly into yours. When I stared into his steely glare, the only word that came to mind was "commit." Over and over I heard the word "commit" repeat in my mind.

I bought the painting and now have it displayed on the wall right as you walk into my office.

I then gave a Commander's Call (which is when I gather my entire squadron into one auditorium) regarding the painting. I explained that George Washington was extending the quill to each of us, to sign The Constitution alongside his name. Much like The Declaration of Independence, he wanted a physical sign of commitment for the bold act they were about to commit. He didn't just want cheap spoken words or a conditional agreement. He wanted full commitment to the words on the paper.

I went on to explain that every time we raise our right hand and swear the "Oath of Enlistment" as enlisted members or the "Oath of Office" as an officer, we are signing our names alongside his. The words we repeat, "I swear to support and defend The Constitution of the United States..." and the signature we sign on our enlistment paperwork, is as good as taking the quill from Washington's hand, dipping it into the inkwell, and boldly signing your name next to his. It requires commitment. It requires knowing the weight that is now on your shoulders. I printed out copies of the painting for everyone to keep and asked them to sign their names at the bottom of the painting. This eventually became the cover page for my "Commander's Packet" that I review with every new member as they join my squadron, but I will explain that in a later chapter.

From then on, the majority of the oaths I would administer were conducted in my office, in front of the American flag and this painting; explaining to the member who was either taking the oath for the first time or renewing their enlistment, the significance of the oath they were about to take.

If the American Revolutionary War hadn't gone the way it did, George Washington could've been written into the history books as a traitor and tyrant, and we'd maybe still have smart sounding accents. Instead, we remain committed to the cause of freedom, like all who have gone after him and sacrificed to preserve those freedoms. We need those types of leaders again in America! In every sector!

Here is the bottom line: Leaders, it's your fault if your organization is good or bad. You need to be committed to your people and work tirelessly for them. You need to be devoted to serving them for the right reasons and hold them to the same high standards you hold yourself. Find your purpose, your passion, and your potential. Then...don't be afraid to lead with your heart and gut.

Chapter 4

If You're Scared, Step Aside

My first week after assuming command of the 243d ATCS, the Deputy Group Commander (my boss's "number two") took me out to a Military Affairs Committee luncheon. He wanted to introduce me to some of the city leadership in Cheyenne, as well as the other military leaders from both the RegAF AFB in town (F.E. Warren AFB) and the Army National Guard (ARNG).

[Note: Cheyenne houses three military entities: RegAF, ANG, and ARNG.]

On the drive to the luncheon, the Deputy was coaching me on how to approach the first few months of command and assured me that I wouldn't need to be making any big decisions within that time. This is a common recommendation: "Don't change anything too soon." I usually prescribe to the "If it ain't broke, don't fix it" mentality, so this type of coaching usually wouldn't have had me questioning if that was the right approach. However, I inherited a broken squadron...

The 243d ATCS was known as the "redheaded stepchild" on base. For years it had been peppered with alcohol-related incidents, investigations, selfish leadership, and a reputation for not being a team player with the rest of the units on base. Most recently, and why they were so desperate to get a new commander (another reason I got lucky landing the position) was that the last commander was relieved of command for loss of trust and confidence. In fact, if you took the previous six 243d ATCS commanders and interim commanders and averaged out their terms, the average "life expectancy" of a 243d ATCS commander was less than fifteen months.

So that's the situation I stepped into. It was a bit of a mess, and they had a Second Lieutenant (O-1) holding things down until I arrived on station. Remember, this is a squadron of just under one hundred Airmen and there are only two officers in the entire squadron: the Commander and the Director of Operations (the "number two"). All others are enlisted Airmen and DoD civilian employees.

So I had my work cut out for me, and remember, I was just coming off an OTS Instructor tour where my flights were squared away, high-and-tight.

When I arrived at the luncheon, the Deputy started taking me around to shake hands and make introductions. This is where I first met our state's two-star General (The Adjutant General or "TAG"; this position works directly with/for the Governor of the state and is in charge of both the Army and Air National Guard within the state). I was also introduced to the full-bird Colonel (O-6) Wing Commander of F.E. Warren AFB (one of only three high-profile strategic missile bases in the U.S.).

When the Deputy introduced me to the Colonel as the new ATCS Commander, his eyes narrowed, as he looked at my commander badge on my chest, and then over to my captain rank. It just didn't add up and you could see it on his face. He asked a few questions to ensure he understood correctly, but once it computed in his brain that this was legit and that I truly was a squadron commander, his demeanor changed. He looked me straight in the eyes with the Deputy standing right next to me and said, "Other leaders will tell you to wait ninety days after you assume command before making any big changes or decisions. Fuck that. Make decisions and change what you see needs fixed. Don't waste any time."

Now the narrowing of eyes was reversed as the Deputy's face changed, since this was exactly opposite of what he had just coached me on the drive over. The next three years were filled with similar glances at my commander badge and captain rank, as well as moments I found hilarious such as this one.

Others may have been scared to step into such a beatdown organization, but having this Wing Commander acknowledge the position I was in, respect the challenges ahead, and then charge me with taking control as if I were a seasoned Lieutenant Colonel, invigorated me. That was all I needed.

If you're scared of making tough changes, step aside.

◆━━━━━━━━━━━━━◆

Although I changed my approach from what I used at OTS as an Instructor, the principles I applied remained the same. I led with my heart (love) and my gut (logic). Love and Logic. Love and Logic. These are principles anyone in any organization can apply and are the foundation to my leadership style.

Fast forward to an email I received while on my Air Advisor deployment from our Recruiting and Retention E-8 for the Wing. He asked the leaders on base to help him understand what makes people stay and what makes people leave the Air Force. This was my reply:

"Our 'why' is the main focus I have had the past three years as a Squadron Commander. By sharing my own 'why' and giving examples of others' 'why', I believe the Hawks I now have in my squadron are serving for the right reasons (their own), which makes retention easy for me. They needed a reason to show up and a reason to be excellent. I constantly show them how 'Fired Up!' I am to show up and be excellent. I then asked them to answer the following three questions:

43

'Why are you here? What is your purpose? What is your potential?' I made them spend a day over drill writing their answers down, for only themselves to see.

I told them 'If you can't answer these three questions, then you shouldn't be here.'

How did I get there? I loved them for eighteen months before giving them this assignment. I led from my heart (love) and gut (logic). This is what people want from their leaders, and they see straight through any other shit we try and portray that isn't as authentic. I truly believe retention issues are the faults of Squadron Commanders. It's our job to build a unit that is excellent and desired. Even further, as commanders it's our responsibility to build a team that won't turn and run at the first sign of the enemy approaching. This is only achieved if we are truly brothers and sisters in arms. Love and Logic, my man."

If you yourself can't answer those three questions ("Why are you here? What is your purpose? What is your potential?") then you cannot successfully lead your people. You could manage them, and you could produce results, but that's not why you're reading this book. You want to be superior. You want to be excellent. You want to be the best at what it is you do. If this is the case, then you have to love what you do, love the people you work with and for, and not be afraid to let logic lead your decision-making process. Stop worrying about what you say or do costing you your job. Instead, start asking yourself if you'd lose your people or their passion, if you didn't say or do the things that *needed* to be said and done. They need to know and feel that you will love them, even when they fail.

It truly is like raising children. Your kids know you will never abandon them and that you will always act in their best interest, not your own. If you had to choose between you or your kids eating, you'd sacrifice your food for them. This is the mentality you have to adopt and constantly show your people. Otherwise, it's just a job and you're just a manager.

This requires taking risk. I will dig deeper into risk during Chapter 14, but it will be a theme throughout this book. You have got to stop being scared and start taking some damn risks!

If you're scared of taking risks, step aside.

◆————————————◆

One of the riskiest ancient military maneuvers was the cavalry charge. According to the National Army Museum website, "Cavalry charges might win a battle, but with poor leadership they could end in disaster. The key to success was the impact caused when fast-moving objects hit slow or stationary ones. Many cavalry charges are remembered and celebrated because of the risk involved."

There are a few key points from this quote. One, cavalry charges were disastrous with poor leadership. Why do you think that was the case? Can you imagine charging into battle and your leader isn't at the front of the charge? Or halfway to the enemy, the leader just kind of trails off to the side and hides behind their troops? Of course poor leadership caused disastrous cavalry charges! The leader must be brave and willing to take the first strike. Are you? If not, step aside.

Secondly, the reason cavalry charges were successful also had to do with physics. When a heavy fast-moving object hits a lighter stationary object...you do the math. This means we must always be charging ahead and maintaining forward momentum. Your people are hungry to feel the pride of productivity doing what it was they were hired to do. Give them a reason to charge ahead. Make it exciting to come to work. Reward and celebrate the successes you find them doing individually and collectively. This will ensure the "charge" doesn't lose speed and your peoples' job satisfaction remains high.

Lastly, cavalry charges were remembered for the risk involved. Yeah, no shit. It's easy to watch a cavalry charge in a movie, think it's badass, and even agree with the risk taken. "Yeah, I would've done the same thing if I were in their shoes. That's the type of leader I am." Yet that same person as a department store shift leader finds it difficult to bend the seven-day return policy rule by two days, because they are worried the store manager would get mad at them. We are so worried about breaking some of the stupidest rules, at the cost of the customer! We hide behind a regulation that was likely written by lawyers after one person screwed it up for the rest of us, rather than use the large brain inside our nuggets to make a logical decision. Why do we put diapers on all of our employees, after one bad employee shits their pants? It's because all leaders have a boss and are doing the exact same thing (operating out of fear), and our employees see that, which also makes the employee scared and therefore not empowered to make sound decisions on their own.

Everything is negotiable. It's not black and white, it's gray. Gray is scary, get over it. Gray is your friend, and you should cultivate a culture where your employees also see gray as their friend by showing them that you won't bite their head off for taking a little risk by occasionally making a "gray decision." Instead, you'll provide "top cover" between them and your boss, coach them through any gray decisions that didn't go well and praise them for those that did go well. "Fail forward faster" is what Lieutenant General Michael A. Loh, Director of the Air National Guard (three star General) told me when he coined me in his office at The Pentagon, *and I live it.*

If you're scared of failure, step aside.

◆———————————◆

I think we all could go on for days telling stories of horrible, risk-averse leaders. Although I want to concentrate on successes and positive leadership examples in this book, I will occasionally share some real winners when it comes to risk-averse leaders and handcuffing policies.

One quick story has to do with an HRO department I was working with. I know, right? You all have some solid HRO horror stories in your back pocket, I'm sure. If I ever meet you, let's have some beers and exchange stories. Better yet, if you're reading this book and you work in HR, you can start moving mountains of bureaucracy and restrictive regulations out of the way with the tools learned here.

So there I was...in a meeting with an HRO Lieutenant Colonel (O-5), a Major (O-4) JAG (military attorney), and three other HRO minions. My boss was also present. We were meeting regarding hiring board processes and notifications. They called this meeting because I had recently been the board president over a couple of command section job openings within my squadron, and I was questioning why I wasn't the first to be notified the day HRO had decided to make the official selection result calls to all of the applicants. As the board president I obviously knew who had and hadn't been selected, but the selection paperwork has to be processed by HRO before official notifications could be made. I simply wanted to be made aware the day HRO was making notifications so that I could follow-up with the applicants to provide feedback that afternoon.

Instead, I would sometimes receive an email or call from one of the applicants often days after HRO notifications had been made, asking if they could schedule a time to receive feedback so that they could improve. Not to mention, I wanted to call and congratulate the person we selected and welcome them to the squadron! I didn't like how it looked, since I am big on providing timely feedback to people even if they weren't "my people" yet and had just applied to my squadron.

HRO explained that typically they don't inform commanders of when notifications had been made. Commanders would usually just get "Cc'd" on the same HRO onboarding email sent to the selectee which had the paperwork needed to start inprocessing. Yes, I was used to getting "Cc'd" on these types of emails when I wasn't the board president, but these past couple of positions we boarded were within my immediate Command Staff and therefore I decided to chair the board rather than delegate the hiring process like I usually do for positions outside of my Command Staff. I hadn't realized that board presidents weren't the first to be notified when HRO was making the official notifications. I told them I thought this was a horrible policy since it didn't afford board presidents the opportunity to reach out and provide immediate, timely feedback to the applicants. Instead, I was getting emails from applicants thanking me for the opportunity and asking for feedback when I wasn't even aware notifications had already been made, since the onboarding email wasn't always sent the same day as notifications.

I didn't like how this looked to the applicants. It could come off as not caring about their success since they weren't hired into my organization as a non-select, or that I wasn't excited to have them join our squadron if they were the selectee. When I explained this to HRO they stated that they could start notifying me first if that's what I would like (yes, thank you). It's the next part that drove me to include this story in this chapter. They cautioned me that I needed to be careful when providing feedback, and that most of the time detailed feedback to applicants was discouraged.

I was blown away. How the hell are non-select applicants supposed to improve their interviewing skills? Also, feedback is no good if it's weeks old; most of the specific feedback will have been forgotten. Hence why I wanted to be notified first that selections were being made that day. I pushed back on this, and the JAG present at the meeting emphasized that "the agency" was at risk by providing the non-selects detailed feedback. It could be construed as "insider information" or "unfair advantage" by providing them advice.

Really?! Is this how far we've come as a litigious society, that I can't even provide interview feedback to non-select applicants so that they can improve? All because we are worried about a future non-selectee suing the agency because they think the selectee had received "insider information" from the commander? If this whole ordeal warranted a meeting between seven individuals, can you imagine what other regulations were even more risk-averse?

I relentlessly pushed back. I pointed out that the "insider information" I was giving was how to improve interviewing skills, what books I would recommend reading to better prepare next time, how to square away their uniform more appropriately, and successful interview techniques like bringing a sheet of paper and pen to write down thoughts while questions are being asked. Yeah, this is some real heavy Wall Street insider info shit. I would also allow them to ask questions about the interview questions so that they had a better understanding of ways to approach certain subjects. We never use the same questions on a repeat board for the same position, so this was safe to do.

The JAG and Lieutenant Colonel still persisted that they were afraid this could get me and the agency into trouble down the road.

While looking the Lieutenant Colonel in the eyes, I said, "I am sorry that you run your organization with fear as your guide, but that is not how I run my squadron, nor my life. I will continue to provide detailed and valuable feedback to all of my applicants once the board is over and the selection notifications have been made. Additionally, I don't care what your policy is, from now on I want to be informed the day you are notifying the applicants, so that I can reach out to them that day and provide this feedback."

Needless to say, that O-5 didn't shake my hand when I offered it at the end of that meeting, but guess what?... I now get notified as requested, and I haven't been sued for providing post-board feedback to applicants. Go figure!

Listen, relationships matter *big-time* in every sector of leadership. Strong networking built with trust and collaboration as the foundation, is how good business gets done. So I'm not promoting being a battering ram all of the time and flippantly burning bridges. However, if you're under the illusion that everyone is going to like you as a leader, then you're lying to yourself and will be very offended when you receive anonymous negative feedback on surveys.

Although I don't agree with much of the autocratic leadership style of the late Steve Jobs, I do appreciate his quote regarding this matter, "If you want to make everyone happy, don't be a leader. Sell ice cream."

If you start making decisions based on the impact to your "approval rating", then you're just another unauthentic politician. Authentic and effective leaders, on the other hand, must sometimes make unpopular decisions and hold uncomfortable conversations.

If you're scared of occasionally making a few people unhappy, step aside.

•◆————————————◆•

I really didn't want to bring up COVID-19, but there are just too many examples of leading out of fear. Just like HRO horror stories, we've all got our COVID-19 stories of stupidity. However, I feel compelled to highlight the "bell curve" of fear-driven policies that occurred during that disastrous time. Mainly so that as leaders we can prevent ourselves from overreacting when one member of the herd jumps at something that startled them from the brush in the distance...which could set off an unnecessary stampede.

The ridiculousness started out slow, then rapidly ballooned out of control, and then after, coming down from our insanity...it was suddenly over. Now we're on the other side of that bubble looking back sheepishly at our overreaction. There needed to be precautions taken, so I am not advocating recklessness (see "Chapter 14: Relentless, Rebellious, & Rogue... v Reckless"). It was a new virus and we needed to prevent deaths. However, it quickly became politicized and other factors that had nothing to do with preventing deaths started clouding the actual level of threat and preventative measures needing to be taken. As leaders, we need to avoid getting sucked into the undertow of escalation of commitment.

Escalation of commitment is a product of compounding bad ideas on top of more bad ideas, because you don't want something to fail, and not being willing to "shoot the lame horse" because you think it's too far to walk back on your own. So instead you ride further into the desert to eventually kill both you and the horse. Logic was often not part of the COVID-19 decision making matrix; instead it was a tablespoon of science mixed with a few hundred gallons of fear. Data should not be ignored, but don't let fear-driven data be the horse you ride deep into the desert when your gut is telling you to shoot it and walk back to the land of logic. That's all I'll say about COVID-19 and the escalation of commitment.

•◆———————————◆•

To close out this chapter on leading scared, I need to breach the topic of constantly being under scrutiny as leaders for what we say, as well as being made to apologize for speaking freely.

I dedicated this book to the freedom of expression, because I have visited and lived in countries that don't have that liberty. Most recently while on my Air Advisor deployment one of the Qatari officers I became very close with asked me what it was that I loved most about being an American. I told him the freedom of expression. He smiled ear-to-ear, gave me a high-five, and said, "That's what I would say too if I were American." This is the same Muslim that said when he gets to heaven, he will be permitted to finally get neck tattoos and drink whiskey while smoking a cigar. Love it.

However, I'm not going to wait until heaven to exercise my freedom of expression as an American, without apology.

Yes, you must know your audience, but even then, your words will be taken out of context and used against you. Hell, this entire book is a rogue move on my part. I am still four years away from the minimum years of service for retirement, which would act as a rip cord to pull in case it looks like I could get fired for the things I'm saying (I plan on going well past twenty by the way). That's a large part of why I'm not waiting until I'm "safe" to publish without worrying that it will affect my pension. I would be a hypocrite of everything I'm stating in this book if I did that. By the time retired Generals publish books, their information is sometimes already outdated and can be out of touch with the true issues occurring "in the trenches" of frontline leadership. Conversely, this is a leadership book for *today's* young leaders to start applying *now*, and it needs to be published immediately so that as American leaders we can secure the future of our country economically, politically, and militarily.

Almost weekly we see instances of people saying what they really mean, and then the next day they are on social media showing their true weakness by apologizing and caving to the pressure of the press, or their sponsors, or their employer.

Why are comedians the last true sentinels of freedom of speech and expression? I don't see their venues being canceled or the number of people attending their shows dropping, just because they are *extremely* politically incorrect. Isn't that ironic? As a public we anonymously throw stones on social media and nod in agreement at the television reporting another apology, yet then turn around and buy tickets to the comedians or movies that portray the exact opposite.

America: stop being hypocrites and start protecting the freedom of expression by not bowing down to what we have been *nurtured* to believe. None of us come out of the womb with political correctness as part of our nature. All of that bullshit is nurtured into us over time and is dependent upon our surroundings. So let's take a step back, stop being so offended and sensitive, and appreciate that our forefathers showed us how it's done when they gave Great Britain the middle finger. Like John Hancock, I want to sign my name in big ass letters under the Declaration of Independence. Let it be known that I am telling you to "fuck off." That's what made us a strong country. That's what other countries respect: strength. Pandering and apologizing is weakness. As leaders, we must take back our right to speak from the heart and gut. Stop apologizing.

If you're scared, step aside.

Chapter 5

People First, People Always

A common military phrase is "Mission First, People Always." This is a copout. It's a way for leaders to put an undesirable task ahead of employees' needs, by framing it as if it's out of their control...thus freeing them from being the "bad guy" by blaming it on "the mission." However, you simply can't always take care of the people if you're putting the mission first. Instead, more often than not the "People Always" part takes a backseat and becomes more of "we always...eventually...get to the people...maybe." Therefore, the phrase "Mission First, People Always" contradicts itself.

Conversely, I created a "People First, People Always" approach to leading an organization toward the same goal. It probably sounds a lot like another leadership cliché: "If you take care of your people, the mission will take care of itself." Which isn't a bad spinoff of "People First, People Always." I just prefer my version because it takes "mission" out of the wording altogether, so that it's not a hidden agenda of *why* you take care of your people. Otherwise, *mission* will still be the driving force behind *why* you lead.

Instead, as leaders, we need to find the much deeper meaning of *why* we do what we do. For my chosen career I ask myself "Why do we even have a military?" In my opinion it's so that we can continue to live our daily lives without fear of attack from another country, and therefore continue to enjoy whatever it is that makes us happy: our families, religion, sports, entertainment, ten dollar morning coffee…whatever it is that you enjoy while not giving second thought to your country's safety, *that* is why we have a military. Otherwise, we'd just be waiting for the next global power to take us over, tell us what to do, and how to do it. Personally, I love the United States of America because I can write a book like this without being censored. I also love going to my kids' sporting events and watching them enjoy just *playing* (which I think is one of the most fundamental and enjoyable things about life…playing). There are so many little things we enjoy about our daily lives that can be attributed to the basic freedoms we have secured.

So, if we think the mission of blowing up a bad guy is *why* we are in the military, then we've "drank the Kool-Aid" and just want to blow shit up for the fun of it. I'm not saying I don't like blowing shit up, but that's not why I serve. Those currently serving in the U.S. military make up less than one percent of the total population. Therefore, when it's time to saddle up and actually go to war, the other ninety-nine percent get to stay back and continue to appreciate those things they currently enjoy, without fear of a bomb landing on their homes. That is why Pearl Harbor, and the 9/11 attacks were so jolting as a country. It's the only two times anyone currently living can remember being attacked on our own soil by a foreign entity.

The U.S. is wise to strategically posture our presence globally even during peacetime and has been very successful ensuring that whatever war it is, for whatever reason, is fought outside our borders in someone else's country. You don't have to agree with why we are at war to appreciate this tactic, which keeps ninety-nine percent of the population out of harm's way from the kinetic devastation of war.

When I deployed to Iraq in 2010 as a Senior Airman (E-3) for seven months, I missed the birth of our third child (first daughter) and came home to a four-month-old I had never held before (my wife is truly the strongest person that I know). It was devastating missing that moment and I was homesick every single day of that deployment, but at least my daughter was born without the cloud of fear hanging over her that Iraqi newborns were born under during that same time. Those families and the vast majority of Iraqis that did not wish harm upon the U.S., didn't ask for the war to be fought in their country. Yet it was, and fear riddled their land for decades, and still does in different ways.

This is why I *serve* in the military.

Why I *lead* in the military, is to take care of the people comprising that less than one percent so that they can protect our homeland from attacks on our soil and get back to their loved ones when we have to deploy overseas. The entire reason my position as a commander exists is to ensure my people are mission ready, which requires taking care of the human more than it does the mission.

Therefore, I serve the ninety-nine, and lead the one percent.

Whatever your profession, always remember the humans that it takes to do what you do. Remember the lives, the emotions, the trials, the joys, and the realness of the human beings behind the mission and daily work.

•◆————————————————————◆•

I've had many occasions in my career where I was quickly sobered by the heavy life issues my people were going through. I could write about several such stories, but I think most other humans are able to do the right thing in those moments. I'm not saying those emotionally tasking instances are easy to deal with, but they sometimes present easier paths toward empathizing and seeing others as vulnerable human beings. Rather, I want to share an experience I lean on from when I was an OTS Instructor at Maxwell AFB, Alabama, that helps remind me to see the human within the uniform even when "it's not personal, it's just business."

OTS is "boot camp" for those who have applied to become a

commissioned officer in the Air Force (AF) and is one of three commissioning sources the AF offers (college ROTC and the AF Academy comprising the other two). OTS graduates more officers each year than ROTC and the AF Academy combined. I've been through enlisted boot camp (Basic Military Training – BMT) as well as OTS, and OTS is a different type of difficult than BMT. At OTS, we still yell at you, and it is somewhat physically tasking, but not nearly as much yelling or as many push-ups as enlisted BMT. Instead, we pile on so many expectations at once to see how you will prioritize and execute to maintain success, that it is very mentally and academically challenging. Whether it was papers to write, tests to study for, briefings to give, field exercises to plan and lead, or physical fitness standards to meet, you had more to do in a day than hours to do it. So you learned how to balance.

Some officer trainees arrive already wearing rank. These trainees are usually medical professionals, lawyers, chaplains, etc. The Air Force starts them out higher than a Second Lieutenant because they are harder to recruit, since outside of the military they would likely be getting paid more for their profession. However, in my mind these commissioned trainees were on the same playing field with the non-commissioned trainees not already wearing officer rank, and both groups had to earn that rank by passing the nine-week OTS.

I made this very clear on the first day of training with my sixteen-member flight of mixed commissioned and non-commissioned trainees, through some choice words about everyone being…let's just say, equally useless until graduation. This helped dissolve issues I overheard other flights having each class, which usually consisted of commissioned trainees trying to "pull rank" on their fellow non-commissioned flightmates. In the five classes that I pulled, I never had an issue with this happening within my flights.

Sometimes the old "break 'em down to build 'em up" theory works.

One such commissioned trainee was a Captain Pharmacist. He had an especially rough start to OTS. You see, I was also a captain, so not only did I need him to understand that he was still an ordinary trainee that needed to earn his rank just like everyone else, but since we were technically the same rank, I needed to ensure from the very beginning that he never saw us as equals. Again, keep in mind this is a military accession course, so the intensity and rank structure are very emphasized. This is not how I run my current squadron, mainly because I don't have to since we were all taught from our boot camp experiences about the chain of command and rank structure, but also because dictatorship is contrary to running a healthy organization. It's similar to the parenting analogy I used in Chapter 3; establish the expectations and "rank structure" early on, and then you'll have less issues with the confusion of accountability later.

Back when I taught at OTS (the course changes over time), as Instructors we did not interact with the trainees for the first four days. We had enlisted Military Training Instructors (MTIs, aka: "drill sergeants" ...the ones you think of in the movies, wearing the "Smokey the Bear/State Trooper" campaign hats and yelling at trainees). We had a handful of MTIs that would teach them how to march, wear their uniform, and get to scheduled events on time. As an Instructor, I learned some of my most valuable lessons from the MTIs, and to this day I highly revere their professionalism and expertise above all else when it concerns building an Airman from the ground up.

On the morning of Training Day 5, at 0400 all of the Instructors would gather in preparation for the "Blue Line Ceremony." During this ceremony all trainees (usually around three hundred each class) would be marched from their dorms by the MTIs to then line up shoulder-to-shoulder in front of a long blue LED rope light. The rope light was positioned at the edge of the perfectly manicured parade field located in the middle of the OTS campus. Here, the OTS Squadron Commander would position himself at the reviewing stand podium and give a short speech through a loudspeaker, explaining to the trainees that they were about to officially enter officer training. The commander would boldly decree that if they accepted the challenge, they needed to take one step over the lighted blue line.

I remember my own Blue Line Ceremony when I went through OTS as a trainee. It was a special moment and very symbolic for me crossing from the enlisted corps into the officer corps.

By this time in their OTS experience, the trainees had just four days under their belt of being yelled at by the MTIs and just a little taste of what the next eight weeks would entail. Every class we always had one to three trainees that wouldn't step over that blue line. They would kindly be escorted off the field and sent home that same day. Their hopes of becoming an Air Force officer, terminated.

While the commander was giving their speech, us Instructors (about thirty strong) would slowly and quietly creep up behind the line of trainees in the grass, stopping just feet from their backs, waiting for the key last phrase in the commander's speech: "Your training begins now!" Before the "w" in "now" could finish leaving the commander's lips, all thirty Instructors unleashed absolute havoc as we descended upon all three hundred trainees. It was complete mayhem. The goal was to have our flight leader find their flight members (they were purposely not organized into their flights that morning, so that they were randomly scattered across the parade field), organize them into marching formation, march them to the dorms to obtain their backpacks for the day, and then march them to our classrooms for our "Instructor Welcome."

[Note: The Instructor Welcome was an intense introduction between the sixteen trainees and their Instructor in their assigned classroom. Instructors would outline their expectations for the course. It usually involved the first five to ten minutes with the trainees locked at the position of attention behind their desks, as Instructors would "pressurize the cabin" with "powerful words of encouragement."]

At Blue Line, nobody could understand anything anyone was saying during this chaotic moment on the parade field. Trainees were frantically trying to organize themselves, running into each other like a group of ants whose hill had just been kicked over by a curious child; eyes wide as golf balls, and faces cemented in fear. For the Instructors, it was heaven. Absolute bliss. Watching the disorderly formation of flights preparing to march provided us Instructors a buffet of infractions to bark at. Nothing they did those Blue Line mornings would ever be correct. You are just wrong as a trainee that day. You are wrong no matter what.

Having always reviewed my trainee's names, backgrounds, future jobs, and headshots, I had it all memorized which ones were mine. I would also be sure to ask the MTIs beforehand who they had chosen to march the flight the first four days (usually a prior enlisted trainee who actually knew what they were doing). Then I would purposefully make the highest-ranking commissioned trainee march the flight instead. The looks on the trainees' faces were priceless when in the middle of the Blue Line chaos, after they had somehow managed to assemble all sixteen flight members, I would call that highest ranking trainee's name out and order them to march the flight. You could see the concern and uncertainty on their faces, because they had to follow the marching orders of someone who had never once in their life marched a flight.

This particular class, the Captain Pharmacist was the highest-ranking commissioned trainee. "Captain Jones! Get your ass up here and march this flight to the dorms to pick up your backpacks," I barked. "Let's go! Hurry up! Get your Captain bars up here immediately!" For the next ten minutes of this man's life, he struggled to coherently put four English words together, let alone sound off with the correct marching instructions. He was running the flight into walls, plowing over bushes, stepping off the cement and into the mud, and vectoring them into other flights like a battering ram. It was all we could do as Instructors in the dark Alabama morning mist, to hide our laughter and shaking heads.

The entire time he's struggling to march the flight, I'm yelling in this Captain's ear about how badly he sucks at marching, and life in general. I am openly questioning why the Air Force ever decided to allow him to wear the rank of captain, let alone advise other Airmen on medicines taken from the pharmacy. I ensured by the time we actually somehow arrived at the dorms (which were only a couple blocks away) that this Captain wished he had no rank, wished he had not pursued pharmaceutical, and wished to never be addressed by me ever again.

Once the flight halted in front of the dorms, I riddled off some other lines of disappointment and publicly "fired" him from marching duties. I then called to the front the prior enlisted non-commissioned trainee who I knew had been marching the flight successfully for the past four days, to take over the flight since "apparently wearing rank before graduating OTS *does not* in fact make you a leader." Done deal. Nobody was going to ever question who the real Captain was, and nobody was going to try and "pull rank" on each other within the flight.

The very next day we held the Physical Fitness Baseline (PFB), which is the Air Force's physical fitness test (aka: "PT test") administered at the beginning of OTS. The AF PT test consists of a mile-and-a-half run, one minute of push-ups, and one minute of sit-ups. There are minimums in each category, as well as a minimum overall score that must be obtained. The PFB at OTS was the initial "weed out tool," meaning if you failed, we could immediately send you home. At the very end of OTS is the PFA (Physical Fitness Assessment) which must be passed in order to graduate and commission.

This is where the Marines usually butt in and explain how much harder their PT test is. And this is where I don't argue that point. Each branch has similar, but different PT standards. We also have different mission sets. However, I will never defend the ease of the Air Force PT test. It isn't difficult, and physical wellness is a pillar of life which should be fairly strong if you're serving in any military. It says more about you mentally than it does physically, and the "ninety-nine" that us "one percent" serve deserve to be proud of their servicemembers' physical and mental fortitude. Conversely, our enemies don't want us to be mentally and physically strong, so when you suck at PT, the enemy wins.

This particular Captain Pharmacist failed his PFB. Failed it horribly. The minimum passing overall score is a seventy-five percent. He scored a thirty-two percent.

I had the opportunity to recommend his dismissal from OTS. I could have ruined this man's chance at fulfilling his obligation to the Air Force for paying for his pharmacy schooling, meaning he would have to repay the cost of school. I could've leveraged my positional authority I had already very clearly established the day before and crushed this man's soul. Since you read two paragraphs ago about how strongly I believe in PT, you can imagine how I might have been leaning going into this decision.

I ordered him to my office, he reported in, sat at attention, and was visibly extremely nervous. First, I explained to him the possible consequences upfront. I didn't bullshit him on the severity of his failure. He needed to feel the full weight of this. I then put him at ease and "took off my hardcore Instructor hat."

He initially didn't take my offer to "sit at ease" which means you can sit normal, not have to sit straight up, hands "knifed" on your thighs, your heels together and feet pointed out at a 45-degree angle. He was not about to risk getting yelled at again. So, I ended up showing him non-verbal disarming. I leaned forward, rested my forearms on my thighs, changed the demeanor on my face, and softened my glare as I looked him in the eyes to find the human inside. I consciously blocked out the fact that he was an Officer Trainee, in uniform, sitting at attention in my office. It sounds weird, but I tried to look at him as if I were his parent. Then I said, calling him by his last name without rank, "Jones. Brother. Is everything okay? Besides getting yelled at and failing your PT test today, what else is going on in your life?"

He immediately sensed my genuineness, relaxed to "at ease," and his face showed there was more at play. He opened up about how he got there, meaning his path in life. He talked about always struggling with physical fitness. He divulged other things going on back home that were on his mind and the newness of military life and the training environment. I would occasionally nod, show caring in my eyes, switch my position in my chair to show that I was truly interested in what he was saying, and asked follow-up questions to his comments.

Once we had been talking for about ten minutes, I leaned back toward him and asked, "What do you want from your OTS experience?" He paused, took a deep breath, and answered, "I want to change my life and I want to positively influence others' lives."

I can work with that.

Instead of recommending him for dismissal, I continued to ask questions and had him write his thoughts down while I asked those questions. I lead him to develop his own workout and healthy eating plan. He used what he knew about himself to construct a path toward the "*why*" of his statement: "I want to change my life and I want to positively influence others' lives." No more than thirty minutes after reporting into my office, this human being had lifted himself up to the genuine belief that he could pass OTS and he had crafted his own authentic plan to do so. He had buy-in that he himself had created, and he had a higher purpose for doing it. If he had said he just wanted a second chance so that he could pass his PT test, serve his obligatory years to pay off his student loans, and not make waves…things may have gone differently.

Conversely, I was able to look into this man's eyes, see the human inside, slightly assist with guiding his plan, and then concur with his plan.

I told him I believed in him, and that I was going to give him a chance to prove it at the end of course PFA, but that I was by no means going to hold him accountable to his eight-week workout and dieting plan. Anyone can follow a short-term plan when they have someone yelling at them every day to do it. I needed to see that this was going to last well beyond OTS. I then went back into "Instructor mode," dismissed him, and he saluted and reported out.

My coaching and leading of this particular Captain at OTS didn't stop after this "come to Jesus moment." Just like anything in life, consistency is crucial to sustainability. Every two weeks I had a formal sit down with Captain Jones to check up on his progress, as well as to see how the human inside him was doing. During the weeks of training I would badger him just as much as the others; the beatings didn't stop just because we "had a moment" together. If I would've softened up on him, it would've deteriorated that initial meeting and the goals set therein. Nevertheless, I did also show him and the other trainees my "real side" as we progressed throughout training. I was personable, funny, and even humble at times about my weaknesses and shortcomings. It takes a balanced approach to lead people genuinely and build that trust that you truly do care for them first and foremost. Believe it or not, I wasn't a hard ass all the time. In fact, after the first couple of weeks I was rarely a hard ass. I just held them accountable to the established expectations, and then had a blast teaching them and molding them into future leaders.

Fast forward eight weeks, and this trainee had lost twenty-three pounds, cut over four minutes off his mile-and-a-half run time, and ended up scoring a ninety-four percent on his PFA. Not only that, but he was also beloved by everyone in his flight; voted #1 of 16 by his classmates. He actually became a leader amongst his peers. When graduation day came around, I had the privilege of administering to him the Oath of Office. With both our right arms raised, I stared directly into his eyes as he repeated those solemn words, and I knew he would keep his oath. I had witnessed the human inside him and the true character of this man. Nothing is more rewarding as a leader than watching someone you lead reach potential they never even knew they had.

It has been over four years since that moment, and I still get texts from him updating me on his career as well as pictures of him at 5K races, which he uses as motivating goals to maintain his physical training. He truly changed his own life, long-term. This was not a moment in time for him. This was now a way of life, and he is out there serving his country honorably, and making others better. I was able to see him as a human, not a means to accomplish the mission, and now he is doing the same in his profession.

People First, People Always.

The reason I decided to write about this experience rather than some of the others I have had that might be more "dramatic" or "heavy," is to demonstrate the simplicity of taking care of your people. It really doesn't take much, so long as it is genuine. We can all see straight through a leader who is just "checking the box" of taking care of their people. You can ask your people how they are doing in passing, write poetic emails that state how much you appreciate them, and even throw barbecues at work, but if you don't genuinely care about them, then none of that matters. You have to love them and use logic to guide how you lead them.

I have witnessed "leaders" grumble over allowing someone out of work in the middle of the day to go let the plumber in the house or watch the clock to ensure nobody was leaving ten minutes early. I've seen "leaders" dismiss advice from lower ranking Airmen without giving thought to the goldmine source that the front-line employees are. I've seen "leaders" pass by desks on a Monday morning and not even slow down to ask how someone's weekend was, let alone maybe drop off a cup of coffee or donut at their desk, just because. Yet, when someone's mother unexpectedly dies in a car accident and that "leader" jumps into action to approve bereavement leave, they think they are some kind of hero and gifted people person. Bro, those are the easy ones...the ones that if you *didn't* do the right thing, you'd be a complete ass. It's the smaller things, added up over long periods of time (consistency), that make employees feel as if they truly are first and always. That's a helluva lot harder to do as a leader. You've got to earn the title of "leader," otherwise you're just another jerk manager...and we've got too many of those already.

I loved all of my OTS trainees, even the ones I kicked out. That's why it wasn't hard for me to kick them out. I was doing what was best for them, and the Air Force. However, with Captain Jones I recognized a human that was going to give it everything he had, and so I took a chance on him by giving him a second chance. I made an *exception to policy*, and although I try to live by the "fair and repeatable" motto for personnel decisions, sometimes that's exactly what it is: an exception. Exceptions are okay as a leader, because it shows you are making situationally-specific command decisions, not just carrying out blanket policies written by lawyers to "protect the agency." There would be no agency without the people! If that type of thinking scares you, then you should go work for the DMV and not be a leader of people. It's okay, we still need DMV workers, we just don't need DMV leaders.

When I finally left my tour at OTS as an Instructor, I had racked up the most positive reviews on the student written End of Course Surveys, never had one negative comment, and was awarded OTS Instructor of the Year along with my third AF Commendation Medal. During my time there, I had trainees from other Instructors' flights coming up to me asking if I'd be the one to administer their Oath of Office or be a guest speaker at their graduation ceremonies. I was asked to write articles about OTS for Public Affairs and was featured in numerous Air Force videos.

This was all great and cool to get patted on the back, but when I left OTS, it was the framed official photos of my flights that meant the most; their uniforms squared away, and their faces resolute to lead fearlessly. I had influenced hundreds of future officers. I had the honor of being a small part of their future accolades and accomplishments. I was privileged to learn from them, just as they learned from me. I discovered how to uncover the human in each trainee and provide them tailored leadership to assist in reaching their goals. My time as an OTS Instructor, without me even knowing it, was the best "prep school" for squadron command. I also would've *never* guessed that squadron command would be my next assignment.

⋅◆———————————◆⋅

"I have an open-door policy." This is a common phrase for commanders and other leaders to throw out there. It makes you sound so personable and approachable, when in fact, most commanders have numerous blockades leading up to even getting to their office door. I'm not saying I don't have an office strategically located at the very end of a long

Command Section hallway, with the Administrative Executive, Logistics Planner, Air Traffic Manager, Senior Enlisted Leader, and Director of Operations as potential "linemen protecting the quarterback" from blindside visits from outside agencies. However, my Command Staff knows that when they see one of our own Hawks walking down the hallway, that they have an unimpeded path to my office. Luckily, I truly have the best command staff ever and they all "get it" and take care of their people as well, so each of them also has an open-door policy and are approachable enough to initiate a conversation if they are ever needed.

Usually most of my Hawks stop by the exec's office first to see if the commander is in, but it's not required. She then initiates small talk as she walks them down to my open door, and I stand ready for whatever walks through. No appointment necessary. If the door is open, I'm ready to talk and my exec knows this, so she doesn't have to ask me first. These are actually the best parts of my day. I can honestly say that when I walk from my parking spot through the front doors of my squadron building, I only know about sixty percent of what that day will hold. The other forty percent just comes at me, and that's the fun part of leading; it's like air traffic control, but with people instead of planes. I would *never* turn away someone that walks through my office door when it's open.

My standing desk is facing so that when you walk into my office my left side is toward the door and I am able to quickly half-turn to greet. When someone enters, I pivot my whole body to face them so that I can make eye contact. What signal would it send if I just kept facing my computer screen? Depending on what the person needs to talk about, I will come around and casually sit on the edge of the desk so that they have my full attention. I'm not all creepy about it, I just show them that I'm listening.

If the door needs to shut for confidentiality, then we sit across from each other at my desk or over in the couches area of my office. I pull my military ID from my computer, turning my screen blank so that I'm not distracted by emails coming in. People can tell when you don't give a shit. So give a shit, in every way, including non-verbally.

Additionally, an open-door policy goes both ways. Meaning, I need to be walking through doors during the day, not just expecting others to walk through mine. I should be walking into the workspaces of my people to just sit down and talk, with no agenda...*daily.* Even leaning in the doorway with a coffee mug in hand sends a casual message of wanting to just say hi (but not like the guy off the movie *Office Space*). It may take a while for your people to adjust to the boss coming by with no agenda, because they are likely used to the boss having a request or task associated with your visit. So don't always have a task or an agenda for your visit! You'll be surprised how eventually your people will open up about the real issues they are dealing with, without forcing it.

At least in the military, when we go TDY (Temporary Duty, aka: "business trip") the "real work" occurs afterhours at the hotel lobby bar. No joke. I can't tell you how many "meetings" were held while out of uniform in civilian clothes, at a bar or restaurant, when the "real talk" occurred. So attempt to replicate that same type of open communication for your people to discuss the *real issues* with you, but during the workday. Just remember that "conducting business" isn't the point of your visit. Your casual "coffee lean" in the doorway, or legs crossed and arms behind your head sitting in their breakroom, is to remind both parties that we are all humans, not programmable robots. Use that time to build a relationship of trust by talking about things other than work.

If you're reading this thinking casual conversation time should be considered an official break on your timecard, you need to go work at the DMV. Don't be that guy. Believe me, I can get more done in an eight-hour workday, than a "clock watcher" can get done in a forty-hour workweek...and these casual conversations are built into that high level of productivity. They don't have to be long; just sprinkled throughout the week, and *consistently maintained over time.* That's the recipe for making Minute Rice in fifty-eight seconds.

What does this open-door policy gain you as a leader? It builds trust between you and your people that they know you are truly concerned for them as human beings, while also providing you a much more accurate understanding of what's going on in your organization. From there you can accelerate your productivity and evolve your mission set.

Apply the "Make It Right" policy. Stop treating your adult workers like children! I promise you, if you truly show that you care for your people and don't hamstring them over the small things like allowing them to quickly swap their cars out with their spouse so they can drop the other car at the mechanic, they will give you much more than if you micromanaged them. Believe me, it works. I trust my people to get their work done. If one week they worked thirty-eight hours, then I know they'll "make it right" because in the grand scheme they are working their butts off, and they will bring more than two hours of productivity to the organization by extending them some basic human understanding.

Even if ten percent of my workforce ends up taking advantage of my kindness by being a dirtbag employee and not making it right, I've got ways to deal with that if it becomes a trend, but moreover, how much am I really losing out on when a dirtbag employee works two less hours a week? Meanwhile, I'm getting more out of the other ninety percent that appreciate being treated like an adult.

One recent example I had pass through my email regarding subtly being treated like a child rather than an adult professional, was a letter sent out to the entire state military department from HRO regarding Military Child Month. They prescribed that in honor of the military child, appropriate purple civilian attire could be worn on a specified day, in place of the military uniform. For those not familiar with the armed services, we love wearing the uniform and have great pride and respect for the cloth of our military. However, anytime we have the chance to wear civilian clothes ("civvies") at work, we will do it! We also love growing our beards out over Christmas break, and then shaving it down to all the different forms of a beard/goatee/mustache the day before returning to work. #stolencivilianvalor.

[Note: For military commanders, "civvies day" is a great morale tool to use throughout the year, so go ahead and use your G-series to approve some civvies days.]

What bothered me about this email from HRO were these lines: "Purple attire being worn must be professional and appropriate. If an employee is found wearing inappropriate attire, they will be sent home to change without pay." Really? Did we need to spell out to adult working professionals to be appropriate in what they wear? Especially military professionals. You think we're going to come to work wearing a purple tube top and daisy dukes?

The kicker in this email was the part about *being sent home to change without pay.* Maybe it's just me, but I felt like a child when this email was sent out. How about you just leave it at "wear appropriate purple civilian attire"? Cool! Thanks for the civvies day! That was nice of you, military department leaders.

Instead, "the man" has to keep their thumb on us by throwing a threat to be sent home to change, like you're back in middle school and your shorts didn't pass the "fingertip rule." Take care of your people, by treating them like adults. Then as leaders, be transformational enough to handle one-off situations, for the time someone *does* show up to work wearing a tube top and daisy dukes. Pretty sure you'll find a way to professionally handle that situation without having to frantically click through your emails to find "that one HRO email they sent out for just these types of cases!" If you treat your people like peasants, you'll get peasant work. Conversely, if you treat them like professionals, they'll give you professional work.

How do you put your people first? What are some real examples of how to build that type of rapport and relationship that allows you as a leader to genuinely show them you are putting them first? Some ideas that worked for me included throwing a squadron Super Bowl party at a restaurant after hours. All we had to do was reserve the conference room, everyone paid for their own food and drinks, and families were invited. Not everyone will attend, but that's why you have to plan numerous events throughout the year that will eventually cover an array of interests.

They don't have to be elaborate or expensive. Even when it does cost money, utilize your booster club's morale fund to give back by picking up the tab. If you don't have limits on what your booster club money can be used for, then you're likely not in the military, which means you don't have as many ridiculous fiscal rules preventing you from spending booster funds on let's say...some *bottled water* for an outside event in the summer heat. Yeah, that's one of the rules. In these cases, quit being cheap and spend some of that "leadership money" on your people. Whatever fiscal laws you must abide by, do so, but find a way to give back to your people.

I can't tell you how far a pizza party went midway through the ten-month planning phase for our Triple ACE event. I could tell everyone had been putting in overtime to stay ahead of schedule, so one day I decided to just go buy a ton of pizzas for the entire squadron and take it easy for a day. My exec warned me that she didn't think we had enough money to pay for that many pizzas and still have money for the end of exercise dinner later that year. I didn't care. Our people needed a break *now*, and I wasn't about to skimp on pizzas, sodas, and cookies. Never skimp on food for your people!

It was super gratifying watching my people fill their bellies with as much pizza as they wanted and take a break from the highspeed pace we had been maintaining for months. And guess what? We found ways to raise the necessary money to ensure we had enough for the other dinner we had planned, which once again was a massive feast with plenty of leftovers. It all works out, so long as you're putting your people first.

One of the best events which took a little more planning than an impromptu pizza party, was our Lumberjack Holiday Party 2022. The booster club ran with the idea to have a themed holiday party in our squadron warehouse on base. Invites, food, activities…they were all topnotch. Everyone dressed up as lumberjacks in plaid, some wore fake beards, we built an ax throwing wall in the warehouse (didn't ask permission to do this), log sawing competition, nail hammering challenge, stickpull tournament, and a food/dessert buffet like never before. We also added a white elephant gift exchange, mustache competition, and a raffle. The booster received some donations for the food and raffle prizes from local businesses. We took a "squadron family photo" in our plaid, which the booster club later printed onto two hundred Christmas cards which I used to send "thank you" notes to all the entities that helped make Triple ACE a huge success. I even hand-delivered these Christmas cards to leaders in The Pentagon, which will be talked about later in the book. We had an absolute blast celebrating our successes and building brother/sisterhood.

Speaking of competition, it goes a long way to make your job fun. Ever visit other countries and notice that all kids are the same? Regardless of nationality or language, they all know how to *play* without being told how. They don't even think about it. It's just built into us to want to "play," but as we get older, we become super lame and forget how to create games out of the ordinary. That's why when I was an OTS Instructor I tried to "edutain" (educate and entertain at the same time). "Studies show" that we retain more when we're having fun while learning. Duh.

So while we were learning how to set up a dirt landing zone (airport) in the middle of Wyoming, I decided to make a team competition. Four teams had to set up their part of the runway by surveying, measuring, and marking the landing strip; the most accurate and expeditious team won. I threw PT in there by having them do push-ups at one station, sit-ups at another, and then fireman-carry each other between locations. Their total number of push-up and sit-up reps were part of the scoring matrix. It was a kick in the face physically, especially since

71

it was during the hot summer weather with zero shade, but everyone rose to the occasion to compete. I had bought a sad, forgotten, little-league baseball trophy at the local thrift store and turned it into a cheesy award for the winning team to sign and display in our official squadron trophy case, located in the entry to our squadron building.

On a separate occasion, we had to stay on base for an exercise and get dressed in chemical warfare gear for four days in a row. This usually sucks...and by that, I mean it *always* sucks—gas mask, rubber boots and gloves, multiple heavy layers, helmet, etc. You are sweating like crazy and barely able to move. I knew by the end of the four days that my people were going to want to choke a kitten.

So I planned ahead and had buckets of water balloons ready for an end-of-exercise water balloon fight, followed up with coolers full of premium ice cream bars that I bought using my own money. I knew it was a hit and that I had broken through to a new level of love and trust with my people, when toward the end of the water balloon fight, they all huddled together and then ganged up on me with one last cavalry charge in my direction, zipping water balloons at me like machine guns. I've never been so happy to get pelted by so many water balloons. Success.

Another time was while we were TDY as a squadron to Fort Carson, Colorado. We were there for a few days to all qualify on the handgun and rifle. We could've done this up in Wyoming utilizing our own base's weapons qualification ranges, but who wants to do the same old shit every year? So I charged my planners with finding a venue that could house our entire squadron at night, with an open concept, and an auditorium. Although we did our shooting at the Air Force Academy thirty-five minutes from Fort Carson every day, we stayed at the Army fort because it fit my requirements for after hours. Every night we would get back to the hotel, change, snag food at the base dining facility (the Army calls them "chow halls"). Then we would stop by the liquor store on base on our way back to the base hotel (pulling two school buses full of adults up to the liquor store ten minutes before they closed was a sight...and that manager still hates us).

Once back at the hotel, we all congregated in the auditorium where I would say a few words (A FEW WORDS…more on brevity during "Chapter 11: Meetings Suck, Don't Suck"), and review the next day's expectations. I allowed civvies to be worn and alcohol to be consumed during the evening debrief. Then the fun. We would play "minute-to-win-it" games and laugh our asses off. One of my favorites was tying balloons to our ankles with yarn, and then everyone tried to pop each other's balloons without allowing theirs to get popped. It was pure mayhem. I did have to call a knock-it-off when our largest Airman picked up our tallest Airman, and pinned him against the second story window trying to pop his balloon…

[Note: In my three years in command, I have only had one alcohol related incident on my watch. And it wasn't when an Airman put another Airman through a second story window, because that situation deescalated quick enough that no harm was done. The one I am referring to is when an Airman showed up for work still under the effects of alcohol, even though he had adhered to the twelve-hour "bottle to throttle" rule. I will talk about this later as well. However, aside from that I have had no other alcohol related incidents in my squadron, because I established very clear expectations early on about how we would consume wisely and responsibly. I put a ton of responsibility on the shoulders of my senior leaders to ensure this message was understood without compromise. Believe me, I am not your "let's be buds" commander. Make no mistake about it, I am the commander, I am the boss, and I take these types of matters seriously. I am rattling off a lot of fun things we've done in the past to give back to our people, but I run a tight ship, and my Hawks understand where my line is clearly drawn. My Director of Operations (DO) would sometimes tell others, "Don't mistake my kindness for weakness." I think that mantra coming from the two officers in the squadron is very well understood by all. There is a balance, and although we have a lot of fun in the 243d, my expectations are extremely high, and I will hold you accountable. Even still, when it comes to making a decision to involuntarily discharge someone from the Air Force, I still see them as humans. Although I have discharged numerous Airmen from the Air

Force, I have a reminder in my phone to send them a checkup text a couple times for the first year after they have been back in the civilian life. I want what is best for them, it's just sometimes the Air Force isn't what's best.]

Wow…that was a long "note" …sorry. Back to the balloon popping TDY!

I made sure the nightly games never lasted more than fifteen minutes, because nobody likes "mandatory fun" when all they want to do is have their own time to relax. So once announcements and games were over, everyone was free to do as they pleased. This is where the venue selection is super important. Since we had ensured this hotel also had very large commons areas, it created locations for our Hawks to congregate and not feel excluded.

For the first time since I had taken command of the 243d, I witnessed different career fields socializing with each other, casually over beer. Air traffic controllers and radar maintainers, HVAC and administration, command staff and power production. Suddenly, we started seeing ourselves collectively as Red-tailed Hawks, and not by our individual jobs. All because we did something different than the norm, laughed together, and treated each other like adult professionals.

You've got to show your people you love them. You can't do this by sitting behind your desk, or simply getting the mission done, or even occasionally getting up in front of them to give them lip service. You've also got to balance your ability to show them you're willing to have fun at work, with high expectations that have repercussions for not being met. It also doesn't have to be elaborate. When I first assumed command, I realized my relationship with my command staff was going to be crucial to our success. My DO had a great idea to go snag a slice of pie off base every other Friday before starting the weekend. We started calling it "Friday Pie Day." It was a simple and small event that gave us something to look forward to. It also allowed us a change in scenery for discussions away from the office.

Those were just a few of the smaller ways I've found to put my people first, and always. The bigger life events should be no-brainers, such as getting two of my deployed Hawks home in time for the births of their children. Again, just because we're in the military doesn't mean we can't make exceptions. There will be times when we have to surge or when we simply cannot accommodate. Those should be very rare. When I found out that I was deploying to Iraq and that my wife would deliver our third child halfway through the deployment, I was given the option to have another controller take my place instead. However, there was only one other Airman eligible at my base to go in my place that fit the requirements, and I knew his current family situation wasn't extremely stable. I knew his family would suffer more than mine from him being gone. So you surge. However, most of the time there are ways around extenuating circumstances. Let's find a way to say "yes" to our people and take care of them the way *they* want to be cared for.

I was watching a commercial for the Fairmont Monte Carlo hotel in France. The General Manager stated, "We will say 'yes' to anything the customer wants, so long as it's not illegal." I wonder if he treats his employees the same way? He probably does. However, sometimes as humans we treat strangers better than our own employees. Have you ever noticed you treat your friends better than your own family sometimes? Ever feel God put you in a family, because you wouldn't be friends with those people otherwise? That's a joke (kind of), but truly we often don't have a choice concerning our co-workers and we often treat those outside our work centers better than those within it.

As leaders, we need to take a step back and remember who *our* customers are: *our people.* We are in positions of authority so that we can serve them and protect them from other bad leaders. That is what a true servant leader does. They serve their people and do everything within their legal power to get them what they need in order to be successful. If your people are successful, then you will be successful. If they are happy, you will be happy.

Lastly, I want to touch on the subject of recognizing your people. Each organization likely has different formal recognition programs. Formal programs are good, and leaders should be sure to submit their people for these types of recognition. Remember, everyone is driven by different incentives, and although one person may not be motivated by a paperweight with their name on it or a golden stapler, their resumes will always appreciate the added value. Additionally, some people don't appreciate public recognition and would prefer other methods of saying "thank you." So getting to know your people on that type of level just adds to the sincerity of the recognition you give.

For example, in the military certain leaders (especially commanders) have their own challenge coin (i.e., "commander's coin"). It's a customized metal coin that leaders give out to top performers, for any reason they deem, without official paperwork needed (unlike official military medals which require a citation and routing through headquarters for approval…which are also great ways to recognize your people). One of my Hawks had performed exceptionally well, but I knew they hated to be up in front of others and wouldn't appreciate the public coining (there is an official way to receive the coin, with a "shake, take, salute"). So instead, I walked into his office, explained why I was so proud of him, and then coined him by himself. He proudly displays that coin on his desk.

Another example is casual recognition. My controllers in Cheyenne absolutely love chips and salsa, but not just any chips and salsa. Having spent time getting to know them and casually hanging out together in the control tower, I came to know which chips and salsa they prefer. So every once in a while, I'll pick some up on my way over for one of my visits and spend a couple hours over crew change up in the tower (so that I maximize the number of my Hawks that I can see in a two-hour period; just eating chips and shooting the breeze, no agenda).

Other non-official recognition can be as simple as going up to an employee that you have noticed is crushing it lately, and at lunch tell them to take the rest of the day off because they have been crushing it. Yes, as a Commander you have that authority, so use it more often!

Also, I would hope other non-military organizational leaders have the same authority. If not…do it anyways. Take care of your people.

Even text messages go a long way. "You are absolutely killing it lately! I can't thank you enough for taking extra time to ensure the slideshow was superior today. Hope you know I appreciate you." As simple as that, but you need to be specific. Don't just text or email, "Thanks for what you do." I hate it when leaders put this at the end of a long email…an email usually tasking you to do something, because why would they send a "just because" email? So be genuine and *try* sending "just because" emails and texts to your people without asking for anything!

Know your people. I keep saying this, but when you truly know your people, it increases your effectiveness as a leader in so many ways. There are a few guys in my squadron who love to ride motorcycles, as do I. So instead of just talking about how fun it is, we decided to create a "Hawk Riders" group. A few times during the summer we send out an invite to anyone in the squadron that wants to join us for a ride at lunchtime to grab a bite to eat in town. Even if they don't have a bike, they can meet us at the restaurant. Although we only ever have a handful of us attend, it's yet another opportunity to create a more genuine bond with my people. Add up all of these different ways, and eventually you hit the vast majority of your people with their specific interests.

Probably one of the coolest things I get to do as a Commander, is promote people. Whenever a promotion order arrives in my email inbox, my entire command section knows what just happened, because I yell "Fired Up! Wahooo!!" all the way down the hallway. Since I am the first to be notified (since I was the approval authority on their promotion recommendation but have to wait for the order to be created), I get to think of creative ways of notifying and promoting the Airman. As part of my effort to create a solid supervisor/subordinate relationship, I first notify the supervisor and together we come up with an idea on how to notify the member.

Although it's not always this elaborate, depending on the Airman and what they might like, I get pretty creative. One time one of our traditional guardsmen (the Airmen that come in one weekend a month) was working full-time at a local tire shop. I called the shop to ensure he was working that day, and then myself, his supervisor, and superintendent all showed up unannounced to the tire shop. He was surprised to say the least. I informed him that he had been promoted, and then we both proceeded to stand atop a pile of tires while I read the promotion order and pinned on his new rank. His co-workers loved it, and he was smiling from ear-to-ear. He will never forget that promotion. Or another time when I gathered a group of us outside in the snow, had everyone make snowballs, and then called the Airman I was promoting outside. We all congratulated him by peppering him with snowballs, then the promotion order was read as we stood in the freezing weather, after which we dropped down and made snow airmen in the parking lot. Or the time one of our "PT stud" Airmen and I did one hundred push-ups (twenty per stripe), together in the middle of the ATC tower cab. We were able to have fun, while maintaining respect for the promotion and rank associated.

People First, People Always. That's why we exist in any organization, because it's run by humans, remember? Recognize their humanity and get to know them on a deeper level. Yes, we still have a mission to accomplish and since we are the leader, we must ensure it does not fail. However, the mission will never succeed, if we fail our people.

Chapter 6

You're Either Coaching It,
or Allowing It

Recently I was in the checkout line and watched as a six-year-old boy reached for a container full of suckers. He accidentally pulled the whole container off the shelf and spilled suckers all over the ground. His mom was standing right next to him. It was a simple innocent accident. However, he proceeded to sift through the suckers on the ground until he found the one he wanted, and then handed it to his mother. I waited to see what the mom was going to do about the spilled suckers, let alone the fact that he didn't even ask first if he could have one. She motioned to him that he needed to pick up the dozen or so suckers that he spilt. He reluctantly bent down and picked a few off the ground and threw them on the shelf, not even placing them back in the original container. The mom motioned that he needed to pick up more. He begrudgingly picked up just one more sucker, and then proceeded to kick the rest under the checkout display. His mom huffed, and then bent down and picked the rest up, as the child watched. Once she had put them all back, she grabbed the child's hand and then placed the sucker he had chosen onto the conveyor belt to purchase.

What did that mother just teach that child? How can she expect her child to do what she says next time she gives him instructions? Let's say she hadn't said anything to begin with and just walked away. What message would that be sending?

As leaders, everything we do and say is being monitored by those we lead. They notice when we don't hold someone accountable to a deadline that we publicly set in a staff meeting, just as much as they pay attention to how you walked past a piece of trash in the hallway at work when you thought nobody was watching. One required action, and the other inaction. Yet, both send just as strong of a message.

When you do these things enough times, you dig yourself into a hypocritical hole that is nearly impossible to get out of when you finally decide it's time to actually hold someone to a deadline or require work centers to be orderly. The last thing a leader should be, is a hypocrite. Yet, without even knowing it, you can send hypocritical messages on a daily basis. That is why it is so important to understand the phrase, "You're either coaching it, or allowing it."

What that mother should have done was first not overreact (which she didn't). Simply acknowledge that what the kid did was an accident, and then provide the expectation to clean it up. "Oops, that's okay, Timmy, accidents happen, but next time you need to ask if you can have a sucker." She didn't do this; she didn't provide a clear expectation: "All right, Timmy, go ahead and pick those up and put them where they belong since you made the mess." At the first sign of resistance, even non-verbal resistance: "Timmy, pick those suckers up now." If this still doesn't work, you grab his hand and physically help him pick up every sucker, and then you make him pick weeds out of your grass for an hour immediately upon getting home (like really).

You surely don't do what this mom did, and allow poor behavior to be rewarded, or worse…actually do the job for him. It's that simple. Yet, in both parenting and leadership we see burnout from idle threats of accountability. Nobody wants to actually carry out the enforcement of accountability, because it's tiring over time. Well guess what, being a damn good leader *is* tiring! You've always got to be on your game and demonstrate these same expectations by holding yourself accountable, while not turning a blind eye when you see others compromising the expectations.

A simple example that I ran into at the very beginning of command was adherence to the Air Force writing standards (the governing document is called "The Tongue and Quill"). Having just completed a tour at OTS as an Instructor, my knowledge of the Tongue and Quill and being able to spot errors in writing was hypersensitive. Since we had to teach it and also grade hundreds of papers, my eye had been trained, for better or worse (sometimes it's a curse). When my executive assistant emailed me her first Memorandum For Record (MFR – which is the standard Air Force writing template), I realized we had some work to do getting the entire squadron up to speed on writing standards.

Let me clarify, I am not one to let the process get in the way of progress. I am also not advocating for sending things back because of a punctuation error, when doing so would create an unnecessary logjam of paperwork (which is one of my biggest complaints about government processes). However, in order to prevent the actual important paperwork from taking longer to process because it ends up getting sent back by the receiver for more serious errors, I wanted the 243d to get the small things correct when it came to writing. Attention to detail.

More than ever before I believe how we write our memorandums, emails, presentation slides, and even texts, says volumes about us. We use these mediums hourly! Since most of these mediums I just mentioned have been replaced with slang and casual written communication (which is being more widely accepted), it gives professionals a chance to further stand out by maintaining a higher level of written communication with efficient and organized methods.

So instead of signing MFRs sent to me with errors (allowing it), I spent many hours over the course of many months teaching my executive assistant (coaching it). This meant having to sit with her, review the Tongue and Quill, and watch her make the corrections. This wasn't as time consuming as it sounds, it just took consistency over a long period of time (finding a pattern here?). Please understand this: coaching doesn't always mean making corrections! In fact, praising the correct behavior is proven more effective (even in the training of animals) than criticizing poor behavior. So when she would send me an MFR with fewer mistakes than the time before, I would praise her! I would try to find the good in the progress being made and highlight it. Eventually, perfect MFRs became the norm, and I would continue to praise her!

Once she had proven proficiency in this area, I entrusted her to be my "gatekeeper" for everything requiring my signature. Three years later and *she* is now showing me things in the Tongue and Quill I never knew about (and it's over three hundred pages long). *She* is now teaching and coaching others that send her subpar work, and now her job is much easier as the entire squadron has gotten better at writing. We were able to elevate our level of professionalism, while also protecting my signature integrity on the dozens of documents I sign a month. All it took was a little coaching, sprinkled consistently over time.

I could have allowed error-laden MFRs to continue to cross my desk, and either signed them anyways (making my organization look unprofessional and possibly signing something with incorrect information on it), or just quickly fixed the mistakes myself (which for one MFR wouldn't take very long but considering all the things I've signed in the past three years, I guarantee making those changes myself would've added up well past the time it took over the course of a few months to coach her).

Yes, it is hard, yes it takes time, yes it requires not letting yourself get burnout and giving up and allowing the undesired behavior. Nevertheless, it pays off! Just like parenting consistently the past sixteen years has paid off for my wife and me. I would bet my house on the fact that all of my kids would pick up the suckers themselves (after having asked if they could have one first), and place them back in the proper location, without any prompting from me. I also guarantee this translates to other parts of their lives. Consistency pays off.

"Execute a prisoner and feed the troops." One of the best pieces of advice I received that helped me start my command tour with "coaching it," came while I was an OTS Instructor and had just been notified of my selection for the commander position in Cheyenne. It came from the Director of Operations (the #2) at our OTS squadron, who is a man I greatly respect. We are cut from the same cloth and immediately knew we were going to get along after our first trainee wake-up at 0430 when we "ran the hallway" together (assigned to the same dorm hallway to "bring the wrath").

He himself had been the rare breed of Captain Commander, but of a Security Forces Squadron. So when he found out I had been selected for command, he also became one of my most relevant mentors. We spent the next few months before I left OTS in many discussions about command and leadership theories. Of all the advice he imparted, the one that stuck with me the most was, "On your first day in command, you need to execute a prisoner and feed the troops."

This means I needed to set the tone right away for the rest of my command. The tone would be one of high expectations and care for the people. In order to ensure this was understood, he recommended I find the "low hanging fruit" (aka: someone who had been given numerous chances and should've been fired long ago), while at the same time finding that thing the entire squadron had been wanting for some time, but previous commanders had never gotten them. I couldn't have landed into a more perfect scenario. Within the first couple of weeks, I was able to carry out this piece of advice.

One of the Airmen had just failed their fourth PT test as I arrived. The regulations state that the commander can give them a fifth chance or discharge them. Personally, I think four failures is extremely generous of the Air Force, so this was an easy decision for me. I discharged this Airman and ensured I did so in the most formal manner; with his supervisor, the First Sergeant, our Chief, and my DO all present while he stood at attention, and I read the discharge order.

[Note: Never devalue the tool of formal discipline, especially in the military. The tools we have at our disposal are intended to correct behavior and outline the path to be successful moving forward. When we don't have them report in, stand at attention, read the entire document specifying their stumble, and then put them at ease and speak to them about remediation, we are destroying the pivotal opportunity to truly help someone change for good. Discharges are different. By then, numerous remedial actions have been attempted and it's time to let them go. We must be guardians of the uniform and ensure the less than one percent are truly representing the other ninety-nine with honor.]

The story of this Airman being discharged so swiftly and without remorse, traveled quickly. The troops immediately understood I didn't believe in multiple chances and that I took PT seriously. (Fast forward a year later and we had zero PT failures and more than half of the squadron scoring above a ninety percent on their PT tests).

Then I was briefed on the need for cold weather gear. If you're not familiar with Cheyenne, Wyoming…it is where God decided to put the majority of his wind and much of his icy snow…yet humans still decided to settle there in 1867 and they still fight His wrath to this day. The previous commander had decided to spend the budget on other things besides cold weather gear. So, I "fed the troops" by buying them their much overdue cold weather gear and told my Resource Advisor ("budget guy") that we'd find a way to make all the other expenses that year work out. And guess what? It worked out.

The act of buying my people their long overdue cold weather gear showed them that I cared for their well-being. My mentor's advice was already paying off. "Execute a prisoner and feed the troops."

Another tool I started using after commanding for six months was a

"Commander's Welcome Packet." I know, it sounds horribly boring and a lot like homework. I strive for brevity and originality in almost everything I do, so putting together some packet to hand every new member that joined my squadron doesn't appear to fit my mantra. However, I needed a "quality control measure" that I knew would assist me in creating a standardized approach to establishing my expectations…with *every single member* of my unit. Since I didn't create it until six months in, I had to give the packets and my expectations briefing to the entire squadron en masse (our collective "come to Jesus moment" which set things straight that had been out of line for years before I arrived). Then I would schedule every new member I received from there on out a time for us to meet one-on-one in my office as I took the time to explain my expectations and intent.

The commander's packet also established a baseline expectation management tool from which I conduct many of my "ETS talks" (Expiration Term of Service, aka: when your current enlistment will end). I meet with every one of my Hawks twelve months out from their ETS, and we talk about their goals for the future. These conversations go one of two ways. One, I want to retain the Airman and am asking them to reenlist for a certain number of years. Or two, I am giving the Airman the remaining year on their contract to demonstrate that they have improved enough to warrant my concurrence with their reenlistment request. Either conversation needs to be founded on baseline expectations to measure against, or else it becomes too subjective. These "ETS talks" are extremely valuable and help ensure I am retaining the right people. It also helps my people achieve more, since they have an opportunity to speak with the commander about their future goals.

The cover of my commander's packet has a copy of the George Washington painting that hangs on my office wall (the one I wrote about in Chapter 3). I give them the "commit" speech and then review in detail the meaning of our mission and vision statement which I had revamped, along with what "Commander's Intent" means (more on this in Chapter 13). The page I spend the most time on is the memorization page. Yeah, I know, again this sounds painful (and maybe it is for each member I sit with), but it gives me a platform from which to passionately convey how this squadron is run and what I expect of each member. It's less about the words I'm saying, and more about the energy and sincerity by which I convey the message, that I hope resonates with the new member. I'd like to think you either leave my office 'Fired Up!' to have joined the 243d or are already plotting ways to transfer…either case is a win in my book. I didn't ask you to join this team, bro, and if hard work and winning intimidate you, then there are plenty of other squadrons I can expedite your transfer to that love avoiding work and losing.

It isn't a lot to memorize, just five items:

1. Air Force Core Values:

   Integrity First. Service Before Self. Excellence In All We Do.

2. 243d ATCS Mission & Vision:

   Mission: Deploy and Employ ATC Services Worldwide

   Vision: Resurrect Raw ATC, While Modernizing Mobile Lethality

3. Three Principles of Unified Power:

   1. Alignment

   2. Action

   3. Accountability

4. ACE & MCA:

   ACE: Agile Combat Employment

   MCA: Multi-Capable Airmen

5. Commander's #1 Need:

   Mission Ready Minutemen

If each of my Hawks couldn't *at least* repeat these five items from memory and speak to each of them, then I was failing as a commander. I try to keep it simple, and this was the best tool I found to create a foundation for standardized understanding of the "what, why, and how" of the 243d Air Traffic Control Squadron.

Since I give this packet and spend time with each new member one-on-one, I have given this expectations brief more times than I could recall. However, I guarantee it is given with passion and genuine ownership, every time.

I won't review each item, but I would like to cover the "Three Principles of Unified Power" or "Triple A." You'll find I like groups of three. My ATC operating initials are "HH" pronounced "Double H," but many nickname me "Triple H" since there are actually three "Hs" in my last name. Most of the lessons I give follow a three-topic pattern. We named our homegrown exercise "Triple ACE" because we were demonstrating and proving three Agile Combat Employment elements. There are three elements to my explanation of Commander's Intent. All of my workouts are in circuits of three…and so on. So "Triple A" for the "Three Principles of Unified Power" confirms my OCD tendencies (as does the fact that my command staff often messes with my office setup ever so slightly while I'm away at meetings, just to make my eye twitch). Really, it's because I have a small brain and can't remember much after being given three things at a time (except for airplanes, so don't freak out).

I created the "Three Principles of Unified Power" after contemplating what it takes to leverage the human potential of an organization in order to produce unified strength. I had seen it with sports teams that were successful; I had witnessed it in politics with certain coalitions; and I had seen its success at OTS while I built officers sixteen trainees at a time.

First, **ALIGNMENT**. This has to be first. Otherwise, great ideas and energy will be spent in multiple different directions, integrity of the team will be compromised, and the leader won't have a standard to measure against. A team must first be aligned with their purpose and strategy, before anything else. Since I was prior enlisted, I went through BMT at Lackland AFB, Texas. BMT is a perfect example of alignment in its rawest form. Although I also graduated OTS and then four years later taught there, it's different in many ways. At enlisted BMT you are mentally, physically, and emotionally torn down to then be built back into what the Air Force expects you to be. They help you align with the core values, learn discipline, and pound into your soul the gravity of the call to arms.

By the time you graduate it is physically evident by those watching the graduation parade from the review stands, that you are aligned. You are crisp in your marching, your uniform is high-and-tight, and your demeanor is professional. This is alignment. However, the types of intense tactics used at BMT are not realistic in an "operational" squadron. The same *principles* apply, but how you carry them out looks different and requires greater finesse. Instead of using a battering ram to align a bunch of knucklehead trainees, you are using surgical pliers to pull things out of people…kind of like the game "Operation."

For me, the first step was aligning everyone under the same mission and vision. Using the documents governing my squadron's responsibility to combatant commanders (the top Generals in DC), I formed the most basic mission and vision statements to help align our purpose. I also made every one of my Hawks read Air Force Instruction 1-1 "Air Force Culture and Standards" to remind them of their basic duties and responsibilities. I then gave them monthly commander's charges in person (I still do every month), which ensure we adjust to the changing dynamic of the future war, while also maintaining that consistent reminder of my expectations.

Second, **ACTION**. A commander's packet, monthly commander's charges, and well delivered speeches are great catalysts for aligning your people. However, without action they are simply words. I'll state it again in Chapter 11, but one of my favorite quotes from my commander's course was "What you *do* out there, matters more than what you *say* in here." Words are cheap. Words are temporarily motivating (much like this book may temporarily motivate you). However, *action* needs to be taken in order to make those words truth. This is why at the bottom of the memorization page I put a "self-reflection" section with three questions: *"Why are you here? What is your purpose? What is your potential?"* Responding to these three questions will naturally require verbs to answer, if you are truly trying to answer them honestly. I order that all of my new Hawks spend time alone with these questions and write their answers down for only themselves to see. I then explain that the answers to these questions contain the action we need as an organization, to start gaining unified power.

Lastly, **ACCOUNTABILITY**. What if the alignment speech hits just right and members' written verbs are actually being carried out? Is that all there is to ensuring your unit is reaching its unified power potential? A better question might be, what if the actions don't line up with the expectations? You still have someone trying to act on the expectations you set, yet they might be a couple degrees off vector. I am an air traffic controller, so I use headings and altitudes to keep thousands of lives and millions of dollars' worth of airplanes from colliding each day (at least I used to...now I control a couple times a month to stay just rusty enough for my guys to put the real controllers next to me while I relive my glory days). That pilot may be singing the company slogan while actively flying the plane as they were taught, however, with an incorrect vector they could end up at the wrong destination, or worse, their safety could become compromised. Therefore, we need to hold our people accountable to the instructions and reward them for when they are doing the right thing.

In the ATC example, when I give a pilot a control instruction, they are required to read it back verbatim. If I don't listen for and acknowledge that they read back the correct heading, that's my fault and I must reach back out on the radio and repeat the correct heading. Or if the pilot just acknowledges with "roger" and doesn't read back the heading, it's my job to ask them to verify the heading I just issued them. The same goes for good leaders. *Please please please* do not be micromanagers! That is the last thing I want you to take away from the accountability portion of "Triple A." What I mean is that you "coach it," and if one of your "players" keeps stepping on the "three-point arc" when you tell them to "take a three," you need to coach them on where the three-point line is. Why? Because we want to win the game and most points win, and three points are better than two! What you don't do is grab the ball from them and shoot it yourself, and you also don't make them write a report on the history of the three-point line. Stop micromanaging and start coaching! There is a difference, and the main difference is backing off once you've given the instruction. Empower your people.

When an organization is aligned, takes action on that alignment, and holds each other accountable to the expectations, then they are truly powerful!

◆———————————◆

About a year into my command I started seeing a trend of full-time controllers leaving my unit. This isn't uncommon in the Air Force with air traffic controllers (both ANG and RegAF). The Federal Aviation Administration (FAA) pays controllers a helluva lot more than the Air Force, and therefore many Air Force controllers jump to the FAA after their first enlistment is up. The good sign was that ninety-eight percent of those leaving my force full-time, stayed on as Drill Status Guardsmen (DSGs, aka: the "weekend warriors"). This is one of the advantages of the ANG versus RegAF; we can keep these well-trained controllers as reservists and not completely lose our return on investment, not to mention assist those still wanting to serve their country to have that part-time opportunity.

Every time I met with a member leaving my unit full-time (and I *always* meet with *every* member), I asked them their top three reasons for leaving. I was less concerned when I would be told money, location, and proximity to extended family. Especially when these people were still very dedicated to the military mission of the 243d and wanted to stay on as DSGs. However, once I started hearing a common denominator of the same names being tossed around as contributing to a negative work environment in the control tower, I became concerned.

I do not want to go into the details of the personnel issues we were having in the tower, but we'll just say there were a few controllers who needed a wakeup call to their poor conduct. I decided to order a standdown of the entire 243d ATC system for a day. Commanders will very rarely utilize a standdown; usually only when there is a death in the unit. However, I wanted to send a clear message that the current culture and environment in the tower cab was extremely corrosive and that changes needed to occur asap. Shutting down an air traffic system that is usually operational 365 days a year (which is the case at Cheyenne Airport) is no small request. However, my Air Traffic Manager was able to make it happen with a couple weeks' notice, as we handed our airspace over to Denver Center for a day.

I then held a group meeting with every controller in the 243d ATCS. This is also rare to get all controllers in the same room at the same time, since they work different shifts and days off. Nevertheless, with a little massaging of the schedule, the schedulers were able to make it happen.

I took time to explain the severity of the problem. We were losing good controllers to the FAA, not because of lack of compensation or location, but because of the culture at work. The other factors of pay and location are hard to control, but our work environment is most definitely something we can influence. I made some changes in the middle management, shuffled lines of authority for scheduling controllers' days off and leave requests, and temporarily moved my Air Traffic Manager from the command section hallway, up to the tower to directly oversee my "rehab plan." I then had my DO spend hours each week meeting one-on-one with the controllers who were needing the most improvement. He developed a remediation plan with the assistance of Labor Relations and documented each meeting. There were some demotions, there were some suspensions, and there were some who left because of the changes. It took nearly six months before stability was reached, but it didn't stop there. Follow-up meetings, mandated HRO courses, and some hard appraisal feedback with lower scores that truly reflected actual performance, all had to happen.

Eventually we were able to stop the "bleeding out" of people fleeing the full-time tower, hired new middle managers, and retired out some others. The culture and work environment continued to shift in the right direction, under consistent reinforcement of the standards and desired end state goals. It took months that added up to over a year of hard, focused work on changing the culture, but it paid off. Now we are the healthiest we've ever been.

This is one of those "coaching it" stories that drives home the fact that consistency must be applied over long periods of time. We are still maintaining these cultural fix actions, and we won't stop. We can't let it slip back to the way it was. Luckily, I have a longer timeline as a commander than most (again, because of my rank and ANG status), so I can watch seeds turn into sprouts, and eventually mature. I'm pretty sure most hall of fame coaches didn't rise, conquer, and retire all in three years' time. I can also guarantee those coaches had more seasons where they didn't win the championship, than seasons in which they did. Nevertheless, they never stopped coaching and changing peoples' lives for the better. These are the truths we must hang onto when we realize leadership is a lifelong calling.

I also share this experience to admit that what works on seventy-five percent of the organization, may not work on the other twenty-five percent. The full-time tower controllers only make up a fraction of my nearly one hundred-member squadron. So I must ensure I am being a transformational leader that can adapt and apply different leadership tactics to different parts of my organization. What works in my maintenance section, likely won't work for my controller section.

The point is, as the "head coach" you need to ensure your "assistant coaches" understand your intent and provide them a clear expectation for what "right" looks like. Then empower the "linemen coach, quarterbacks coach, and special teams coach" to each lead their respective sections how they know best. Occasionally, you may need to call the entire team in to "take a knee" and listen to very pointed expectations from the head coach in order to correct misaligned behavior. Don't hesitate to do this, but always ensure you have your leadership team's buy-in first so that they can carry out your intent moving forward. Build your coalition of believers and doers, seek the buy-in of the influencers, and never abdicate command.

•◆———————————————◆•

Regarding coaching your leadership team. For me, my leadership team is my Command Staff consisting of ten personnel including myself. When it comes to hiring boards, I delegate the majority of the hiring process to my Senior Enlisted Leader (SEL), Superintendents, and Air Traffic Manager. They (the board) select an applicant, and I concur by nominating the selectee for hire. However, when it comes time to replace one of my Command Staff personnel, I elect to chair that board. I write specific questions for the interview, I handpick board members from around the base that I know I can trust, and I enter the board with zero pressure to hire anyone. I'd rather wait for the right one than force mediocrity.

When it came time to hire my next Chief of Maintenance, I went against the norm and decided to convert that position into an E-9 Senior Enlisted Leader position. An SEL is the direct advisor to the commander regarding all enlisted issues and should be just as close of an advisor to the commander as the DO (#2 officer). I then realigned my other positions to create the three Superintendent positions (Operations, Maintenance, and Support). It was a bold move that other ATCSs didn't align with, but it has proven extremely valuable in the 243d.

The entire hiring board for the SEL position was unanimous on our next Chief Master Sergeant (E-9). He and I hit it off immediately and he was able to consume large amounts of information quickly, and most importantly understand my intent. No more than three months into his tenure with the 243d, we had a pair of brothers in the squadron test positive on a drug test. Fast forward to their discharge day, both of them standing at attention in my office with their discharge paperwork on my desk, and the office full of the usual leadership present for these types of proceedings.

I must preface, ninety-five percent of the time I have a smile on my face, I am blasting music from my office, I am walking around offices chatting and laughing. I am not a hard ass, and I don't "go OTS Instructor" often. However, I do strongly believe in the old-fashioned military ass chewing, when appropriate. I don't believe in a "gentler, kindler Air Force." That's the bullshit our enemies want us to adopt. Can we at least please keep the military, *our warfighters*, free from the "cupcakes and kid gloves" coaching style?

In the spirit of ass chewing when appropriate, I "went OTS" on a pair of young brothers within my squadron. One, because I knew them well enough to know that the "kinder, gentler" approach had already been taken multiple times by both military leadership, as well as their own parents. If this were my last chance to help them change their lives as they exited the Air Force, I wanted to ensure I had at least attempted to round out their life mentorship. This is what I mean when I say I actually love all of my people, even when I'm giving strong feedback. I saw potential in these two young men, and I wanted them to succeed well after they were discharged (which is also why I decided to issue them a General discharge, rather than Dishonorable).

So I passionately, in full control of myself, went atomic on them. Although my doors were closed, I was told later that day that those in the offices downstairs and on the other side of the building could hear me. Good. These types of stories told around the squadron help drive the understanding that the commander will indeed hold his people accountable.

After I had blasted them for a solid ten minutes, I put them at ease and had them take a seat. As I caught my breath taking my own seat across from them, I asked my SEL, who was standing next to me, if he had anything he'd like to add. Without skipping a beat, he stared at both of them with a face like a pit bull, and calmly yet sternly said, "Take your squadron patches off and put them on the commander's desk right now. You no longer represent what the 243d embodies."

I just about fell back in my chair.

All of my yelling, and I thought what he did was even more powerful.

The sound of these young men slowly, disgracefully ripping the Velcro patches from their uniforms rung louder in my ears than anyone had ever yelled. You could see the disappointment in their eyes as they placed their patches in front of me on my desk.

From here we talked about the formal discharge from the military, but then moved into what they were going to do once they left my office and entered back into the civilian world. What changes were they going to make? How would they capitalize on this pivotal life event so that it became a catalyst for bettering the rest of their lives? It was a very powerful meeting.

I still check in on these brothers as they make good on this experience.

The point of that story had more to do with choosing the right assistant coaches. Sometimes you don't have a choice, especially in the military. However, you are the head coach. It's your job to train your assistants regarding your mission, vision, and intent. After only three months together, my SEL knew exactly what I expected of him and of our people. He fully backed my play, and I respected his approach to achieve the same effect. Do not allow things to just happen in your organization. If good is happening without your constant coaching, even better! Don't be a micromanager, and still coach during these situations by praising and recognizing the positive behavior.

Holding your people accountable is a balance. You must still see them as humans, as I outlined last chapter. I believe reinforcing good behavior rather than pointing out and punishing bad behavior, is much more effective and has longer-term positive outcomes for your organization. So as a leader I charge you to constantly be recognizing that ninety-five percent of your day is filled with people doing amazing things for the unit, and you need to be showing and telling them how much you appreciate them on a daily basis. Even if they aren't totally amazing yet, you can still find what they are doing well and highlight that. The praise must definitely outweigh the correction. Make it a measurable goal if this doesn't come naturally, and tally how many positive comments, texts, or emails you gave that day, and don't think five is enough. We must show our love and appreciation for our people and their work on a daily basis, or we are failing as leaders to recognize the good that is actually occurring around us.

I'm not advocating to be a cupcake leader and sprinkle sunshine on turds. Too much praise all of the time starts to become ingenuine. So really mean it and find specific things to praise.

However, I do want to spend the remainder of this chapter talking about a few of the times I've had to take aggressive corrective action. Leadership after all, is not all unicorns and rainbows. At the end of the day you are the one responsible for the success of your people, and if you recognize a cancer that could infect those that are currently performing excellently, then you have to apply radiation or even remove the cancer surgically.

While at OTS as an Instructor, I had to make a really difficult call, all on my own. There are two graded papers due throughout the course in order to graduate. These papers are difficult to write. Not only are we grading the formatting in accordance with the Tongue and Quill which is a very different writing style guide than most of the trainees used in college, but the content must also be graduate level work. With trainees coming in from all over the country, from different universities, with different writing styles, this can prove very difficult for many to adapt to the Air Force style of writing and test taking.

One particular class I had a very interesting flight dynamic. I had the youngest and oldest trainees in the class of nearly four hundred Officer Trainees (twenty-one- and fifty-one-years old). The fifty-one-year-old had served previously in a different branch for a few years, back in her twenties. During her huge break in service she had become a nurse and then decided to join the Air Force. Usually this would be too old for the military, but her specialty of nursing made her a highly sought after professional, and with her years of experience in the private sector she was able to obtain a waiver. She also came in as a First Lieutenant; one of those direct-commission trainees.

I had very little issues with this fifty-one-year-old trainee throughout the course. She wasn't much of a leader among her peers, but that's not what the Air Force hired her for. However, she struggled with writing. Her first graded paper she earned a fifty-three percent. If she failed the second paper, she would not graduate. There is a lot of pressure put on the trainees, but academics are probably the most challenging portion of Air Force OTS.

There are very strict rules on not getting assistance from your peers for things such as developing study guides or writing papers. You are not even allowed to review each other's papers. We do this so that we know the true individual capacity of each trainee before trusting them to lead a team. The rules are explained numerous times at the beginning, and also before every writing assignment. As Instructors we stress the importance of doing all the work yourself and highlight that they cannot even have another trainee read their paper to provide feedback.

In order to prevent bias, after someone fails a paper and is given remedial training, the second paper is graded by a different Instructor. When I received this trainee's second-attempt paper back and she had scored a ninety-eight percent, I first reviewed the paper myself and sure enough came up with the same score. She had absolutely crushed the second paper and I was so proud. I immediately went into the flight room (classroom) and announced to the entire flight that she had not only passed but had scored a ninety-eight percent! The room cheered and I told her how proud I was of her. She said thank you and then thanked her flight mates for their encouragement and support throughout the course. She went on further to thank one of the trainees in particular, for looking over her paper and providing extremely helpful feedback.

At this moment my face maintained a smile, but my "brain's eyes" bulged out of my head. Did she just admit to having a flight mate help her with her paper? She knows that is a dismissible offense, right?

I played it off and said congrats again and then walked out.

When I arrived at my cubicle, I processed exactly what had just happened. Nobody else knew; just me and my flight, but it was blatant. She admitted to cheating, and you could see it on everyone's faces, especially the trainee that had helped. My first reaction was to run to the commander and tell him everything; let him tell me what I should do and how I should handle this. Then I realized, this is exactly the opposite of what I was charged to do as an Instructor. The commander expected me to make those difficult decisions and make recommendations to him. He was very good at empowering his Instructors and made it very clear that he would support our dismissal decisions, so long as we had the proper cause and evidence.

So here I was, just days before graduation from a nine-week course, debating how I should handle this. I could just keep it between me and the flight, and she would graduate. However, what would this say about my integrity, let alone the message it would send to my flight? Was I going to coach this, or allow it?

I decided to call the trainee into my cubicle that had allegedly helped her. This was a prior enlisted fourteen-year Master Sergeant (E-7). He not only had more years of experience in the Air Force than me at that point, but he was an absolute stud. Voted #1 in his flight by his flight members, week after week. He scored the highest on all tests, led field events with ease, and was not only my nomination for Distinguished

Graduate within my flight, but he was #1 overall in the class of nearly four hundred trainees, which would earn him the Top Graduate award. He was also going to earn the PT Excellence award. Pretty much, his name was already getting engraved on the plaques of every award you could possibly earn at OTS. This had already been decided since it was only days from graduation.

When I called him into my cubicle, he was visibly stunned. I got straight to the point and asked him if he had helped the First Lieutenant with her paper. Without hesitation he admitted to doing so. He explained that she had been asking him for weeks to look it over, and every time he would remind her that this would constitute cheating. Finally, she sent him the draft via email without his permission, and he admitted to opening the attachment. Upon reading the paper, he said it most definitely would've failed again. He said it was horrible and decided to do the wrong thing for the right reason and provided feedback which undoubtedly helped improve her paper.

I asked if he would be willing to put what he said into writing and sign it. He agreed. He then asked what was going to happen. I looked him straight in the eyes and explained that both he and the First Lieutenant could possibly be kicked out of OTS. If so, he would go back to being a Master Sergeant, not having earned his commission, and she would likely not be permitted to come back. I dismissed him, and he reported out of my office.

I then called the First Lieutenant in and asked the same question. She stumbled over her answer. She was sidestepping me and using vague words. Realizing I needed to make this official, I ordered her to write an MFR explaining how she wrote her paper and sign it. Due by close of business that day.

Both trainees did as ordered. When I read the Master Sergeant's MFR, it was exactly as he had verbally admitted to me, without excuse. He owned it completely.

The First Lieutenant, however, wrote a vague MFR that simply outlined how she researched and wrote the paper, almost like "how to make a peanut butter and jelly sandwich." She obviously realized her mistake of thanking the trainee publicly and was now trying to avoid further self-incrimination.

I needed to sleep on this one. I went home that night and kept rolling over different courses of action in my mind, weighing the ethics of each and the balance between justice and mercy.

The next day on my morning run, I found clarity and decided to recommend a Letter of Reprimand (LOR - formal disciplinary paperwork) for the First Lieutenant. She would still graduate OTS, but what this would do is ensure her time in the Air Force was limited to less than two years, since being promoted to Captain with an LOR is nearly impossible, especially for this type of offense.

For the Master Sergeant trainee, since he owned up to the infraction, had made the wrong decision for the right reason, and I knew would take this experience and allow it to mold him into a better officer, I recommended that he also be permitted to graduate. However, I asked that all of his awards be stripped, and he only be allowed to graduate. This meant no Top Grad, no Distinguished Grad, no PT Excellence, and no record of his superior performance at OTS on his record.

This was so hard. I wanted someone else to make the hard decision.

When I presented the situation and my recommendations to the commander, he fully supported. He brought each trainee into his office separately and explained the decisions.

The Master Sergeant was extremely relieved and fully accepted his punishment of being stripped of all honors at graduation.

The First Lieutenant said very little and continued to sidestep, not admitting to any wrongdoing.

The morning of graduation and commissioning I pulled the Master Sergeant into my cubicle. I had saved the gold Second Lieutenant bars I had been pinned with at OTS, and thereafter wore for the two years before pinning on First Lieutenant. I explained to this man that I knew he had an amazingly good heart, and that he would make an extremely great officer. With my eyes watering, I presented him with my Second Lieutenant bars and charged him to wear them with pride, because he had earned them. With tears in his eyes, he accepted. Later that day at the graduation ceremony when his family pinned those bars on his shoulders, I knew I had made the right decision.

I still receive updates from the Master Sergeant (now officer), telling me of his successes and accolades. I am extremely proud of this man.

The First Lieutenant nurse never made it to Captain.

This was a pivotal experience in my officership, because it taught me the value of empowered decision making. I had made recommendations to commanders in the past regarding trainees' careers as an enlisted ATC Trainer, and again as an ATC Instructor at the schoolhouse (my first Instructor assignment). Nevertheless, this one I could've easily kept a secret, I could've "allowed it," or I could've run to the commander and made him decide what to do.

However, I had been charged by an empowering leader to make the tough recommendations. This experience would prove to help guide many other decisions I would later make as a squadron commander. Don't rob your people of these types of opportunities because you're clutched so tightly onto power. Instead, leverage "The Three Principles of Unified Power," and watch how much stronger your organization becomes because you entrusted and distributed that power amongst your people.

Chapter 7

If I Gave You $1M...

The day after Halloween 2021, I started a cross-country drive down to Mississippi. I was enroute to meet eight of my Hawks in Gulfport, MS, for a military exercise called "Southern Lightning Strike." We had participated and controlled during the Southern Strike 2021 (SSTK21) exercise earlier that same year, which is a completely different exercise than Southern *Lightning* Strike 2021 (SLS21) (gotta love the South...bless their hearts).

We had dominated ATC at SSTK21, and word of our capabilities got out, so we received an invite to SLS21 (again, very different exercise, so don't be fooled by the similar names). It was part of my campaign to get our name out there so that we would receive more invitations to Agile Combat Employment (ACE) exercises. It was working and I was 'Fired Up!' for SLS21!

I was driving my own vehicle from Colorado down to meet everyone else from my squadron who were flying into Mississippi. I was doing this because two days after SLS21 ended I was scheduled to attend a two-month training course in Alabama (Squadron Officer School – SOS). Yes, I was a current squadron commander that hadn't had the opportunity to attend SOS yet, which I know sounds wild for those who know what SOS is. The timing just happened to work out that I could attend SLS21, catch an Alabama football game over the weekend (gotta love the South), and then start SOS on Monday.

It was going to take me two days to drive from Colorado to Mississippi. On the morning of the second day of driving (that evening I was supposed to meet my team at the Gulfport Airport and had planned on having hot pizzas and cold beers waiting for them), I set out early from eastern Texas.

As I was approaching the border to Louisiana, my heater went out. Then a few miles later my thermostat light came on. I pulled over, popped the hood, and noticed my coolant level was dry as a bone. So I poured some water into the radiator and carefully limped the car to the nearest exit with an auto parts store. I found the coolant needed, topped it off, and got back on my way. No more than ten miles down the road smoke started billowing out from underneath my hood.

So I took the next exit, parked in a store parking lot, and popped the hood again. This time fluid was spewing all over the back of the engine compartment from a leak in a hose. I searched on my phone for local repair shops and started calling one after another. It was the fifth shop I called that said they could look at it right away. I poured more water in and limped it to the shop. Once I arrived (at some junkyard looking shop off a random backroad), this really nice southern man took a look at it (I was stationed for a cumulative of five years in the deep south, so I learned to truly love the people).

Multiple hoses and a connecting bracket needed to be replaced. He got his parts guy running to pick up the part locally and then started taking the bad parts off. By now I had eaten up any extra time I had built into getting myself checked in and pizzas bought before meeting the crew at the airport. Also, I was supposed to attend the afternoon pre-mission brief with the exercise planners and other participating units' leads.

Luckily, one of my part-time Hawks who works for the FAA in Georgia, was also driving over for the exercise and was going to get there in time to attend in my place. However, I'd be lucky to get there for a late check-in and a very late pre-brief with my own crew (the exercise started the very next morning). I was frustrated that my perfect plan was falling apart.

Five hours after sitting in this guy's shop, making calls to the exercise planners, coordinating my controller from Georgia to attend in my place, and passing word to my Hawks that were currently airborne overhead inbound to Gulfport, I was set to get back on the road. I gave the shop owner a hefty tip, along with a squadron patch and one of my commander's coins. I thanked him for allowing the mission to go on and I sped back onto the freeway.

About two hours later the thermostat light came on again! I took the next exit, checked the coolant...dry as a bone! Obviously, there was more wrong than met the initial eye. By now I'm starting to get worried about the damage to the engine. Not to mention, I am in the middle of nowhere Louisiana.

As I'm writing this, I'm realizing the condensed version doesn't sound too horrible. However, at this point I'm stranded hours between any two legit towns, with a car that I needed to follow on with me to Alabama, and a huge exercise in which the 243d had a budding reputation to nurture. Getting to the exercise late was simply not an option. I was determined to make it to Gulfport that night, along with my car.

So, with the limited signal I had on my phone, I searched for the closest U-Haul: Natchitoches, Louisiana. Yeah, sounds made up (gotta love the South).

However, I still had twenty-five miles to drive to get to the U-Haul, and only half a water bottle left to pour into the radiator, but at this point I didn't care if I burnt my engine out. I was going to make it to that U-Haul, rent a truck and car dolly, and pull my broken ass car down to Gulfport that night!

I drove slowly through back roads (because the U-Haul wasn't right off the freeway, of course), passed a few murder houses, almost hit an LSU fan (which wouldn't have been the worst part of my day), and finally chugged into the U-Haul parking lot. I ran into the office and asked to rent a pickup truck and car dolly. A car dolly they had, but no luck on the pickup truck. Instead, I had to rent a fifteen-foot moving truck to pull the car dolly. Whatever, I'll take it. This was out of my own pocket; no way the government's going to approve this "good idea fairy" on my travel voucher.

I need to point out that I wrote a squadron uniform policy which outlines that any time we travel TDY, business casual would be worn on travel days: long sleeve button-up, tucked into pants, with a belt, and closed toed shoes. Think what you want, but when the Hawks travel places we bring the next level of professionalism, and it's noted.

However, at this particular moment I was regretting my policy letter...as I crawled under the car dolly in the sticky Southern humidity (gotta love the South) and hooked up the chains. I threw my briefcase in the back of the fifteen-foot trailer, just so I felt like I was getting my money's worth (just kidding, but I did jump up into the "Mom's Attic" for old time's sake, just to remember what it was like to be a kid...when you didn't have to solve such problems in life).

I got on the road and started calling repair shops in Gulfport. I pulled into a shop after dark that had stayed open late to receive me (I really *do* love the south), and a few of my Hawks were there in a rental to pick me up and take me to the restaurant where the rest of my people were eating dinner. I got to the restaurant and was super relieved to see everyone. I received the back-brief from the Hawk that attended the pre-mission brief, and then I briefed my Hawks (my wife giggles at the word "brief," and I just used it three times in one sentence), right there in the restaurant with a beer in hand, still in my oily business casual. Success.

The rest of the week went amazingly well. We blew the minds of the exercise planners and controlled like beasts. My Hawks outperformed even my own expectations for them, and the 243d Red-tailed Hawks further proved their ACE ATC capabilities. My car was fixed by the last day of the exercise, and I drove over to Alabama for SOS.

Humans love the path of least resistance and often throw in the towel after a couple of tries. A good excuse is enough to slow-roll anything. One question I sometimes ask when I get a quick "no" or "unable" or any type of excuse that I know is someone just being lazy is, "If I gave you a million dollars to find a way to get it done, do you think you could?" Yes! Of course they could and would! Then it *is* possible and you're just being lazy and would rather not put in the needed effort.

Also, you may not want to assume the risk necessary to get the job done. Risk-averse leaders are as easy to come across as sliced white bread at a barbecue joint in the South; the difference is, one goes well with brisket, and the other is never willing to risk it. Find a way to get to "yes" and accomplish the task!

General Mike Minihan is an officer I appreciate. As I write this book, he is currently the four-star General over Air Mobility Command (AMC). He doesn't mince his words or try to coat things in veiled terms because he's worried about his career. He speaks boldly and from the heart. He says what needs to be said. Recently, he wrote and signed an MFR that he sent to all his AMC Wing Commanders regarding how we are not currently prepared to win the fight against China, but he has built a plan to be ready. He was very clear in his direct orders, and excerpts from his MFR are as follows:

"Go faster…My expectations are high, and these orders are not up for negotiation. Follow them. I will be tough, fair, and loving in my approach to secure victory."

"If you are comfortable in your approach to training, then you are not taking enough risk."

This is a four-star General speaking directly to full-bird O-6 Wing Commanders. He is ordering these leaders to step outside their comfort zone when it comes to training their people for the next war. He is encouraging deliberate risk taking. Hell yeah!

So when I ask my people to get something done that is deliberate and not reckless, I remind them of the reason we exist as an organization. Your people need constant reminders of the importance of their jobs and the significance of doing their jobs excellently. I find it easy to know this for myself, especially being in the military and in an operations squadron. However, depending on the type of organization you lead or are a part of, it may be more difficult to find that importance, but find it you must.

One example might be Public Affairs. These are the professionals that produced our stellar Triple ACE video. That video has helped propel the 243d ATCS into a new CAOS mission set (Combat Airfield Operations Squadron, pronounced "chaos"), much more effectively than a written report could have. It made believers out of Generals, in just six minutes! Now the 243d is being transformed into a modernized weapon system, ready for the next fight! All because a team of Public Affairs Airmen took their jobs seriously and know the important role they play in our national security. Find what makes your organization important, and then consistently remind your people of that "why."

Regarding risk and our Triple ACE Annual Training event (TAAT). Enormous amounts of risk had to be taken in order for this event to go as well as it did. For one, Cheyenne had never certified a hot pit refueling (HPR) location on their airport apron. An HPR is when an aircraft keeps its engines running and gets refueled at the same time. It cuts the refueling process time by nearly three quarters; meaning it only takes a quarter of the time to refuel than if you shut down all the engines. This is extremely significant when it comes to quick-turning aircraft, personnel, and equipment back into the fight.

However, the process of getting an HPR site certified usually takes up to a year, with numerous entities involved. Why? Because risk significantly increases when you refuel while engines are running. By the time the HPR became part of the mission planning for TAAT, we were only five months out from execution. Nevertheless, I charged my team (which now consisted of many outside organizations that would meet weekly for TAAT planning) with making it happen. When someone would doubt, I would encouragingly push back by reassuring them that I knew they could get it done, and that I trusted in their expertise to find a way to expedite the process. I would then tell them it simply had to get done in order for TAAT to be a success. I gave them "shares of the company" to help them gain ownership in the overall accomplishment of the exercise, so that they personally wanted it to happen. It became an intrinsic motivation for them.

Sometimes you can't take no for an answer. Find a way. Let your people know you're not accepting failure on this one.

One month prior to TAAT, we got the HPR site certified. Now the Wyoming ANG forever more has a certified HPR site.

As we were in the early planning phases of TAAT, we had great ideas, but no precedence. So, we just started cold-calling other military units to see if they'd want to participate...at their own cost. I built a pretty solid slide deck presentation for my sales speech and would walk into places with my laptop under my arm, ready to open it up and start selling. When I couldn't meet someone in person, I would meet with them virtually, and show them the slides and sell.

When I didn't have a contact yet, I would call the base recruiting line (always plastered on the front of every base's web page) and act like I dialed the wrong number, then ask for the number I was actually looking for. One example of this was when I wanted to get some fighter jets in on the action. The closest F-16 base was in South Dakota. So I called their recruiting line, played dumb, and when they asked again who I was before giving me the number to the F-16 Fighter Squadron Commander's office, I left my rank off and just stated that I was the 243d Air Traffic Control Squadron Commander out of Cheyenne, Wyoming (since telling them I was a Captain Commander would sound made up). It worked.

I called the F-16 commander's office and started selling. Long story short, after convincing him that I was indeed a commander, we got two F-16s committed to participate in TAAT...on their own dime. As I was working with the lead pilot during planning, I really appreciated her warrior attitude toward making things happen. I explained that we were attempting to prove the ACE concept in every way, and she wanted to do the same: show how light and nimble the "F-16 ACE package" could actually shrink down to if needed. When talking logistics concerning the number of personnel needed on the ground in Cheyenne to support, she said it's normally five. However, later she went back to her Crew Chief and asked what would be the minimum? He replied two, but only during contingency operations. She replied, "Why would we train differently than how we play?" Hell yeah! So they planned on two.

Unfortunately, due to runway construction which temporarily shortened the length of the runways in Cheyenne (which was originally scheduled to be completed before TAAT kicked off but wasn't) the F-16s weren't able to come. However, the majority of the lessons learned had already occurred. Hustle to find a way, and practice how we'll play in the real game. I truly appreciated this fighter pilot's ACE mentality.

Fast forward to the night before STARTEX (start exercise) of TAAT. My entire squadron had convoyed a couple of hours north to the Army camp from which we headquartered out of for the week (where my command cell and I orchestrated the event live from the Tactical Operations Center, or "TOC"). All other players were postured for execution the next morning. Everything was in place. I held my post-dinner squadron brief about how the first day would look, then gave the troops the rest of the evening to relax.

At around 2130 that night, I was in my dorm room reviewing my game plan for the next day, when the Army CH-47 Chinook helicopter pilots from Colorado called me. They informed me that all CH-47s had been grounded nationwide due to mechanical issues (this is like "recalling" a car and requiring they not be driven until all cars of that model were inspected for the same problem). There was simply no way around it, the CH-47s were not going to be able to participate. Which wouldn't have been a huge deal if they were just flying around giving us more traffic in the sky to control, but they were our main mode of transportation out to the other five locations each day!

As a reminder, we were proving a "hub-and-spoke" ACE ATC operation where we "hubbed" out of the same location every day, were infilled to the "spoke" landing locations, and then exfilled at night back to the hub. This was a huge blow to the entire operation.

I quickly gathered my command team and told them the news. Immediately the team kicked into action with an attitude of "no quit" like I've never seen before. They were not going to let this fail. They *would* find a way. I actually took a step back and realized they didn't even need me. They had learned through ten months of planning and not taking "unable" for an answer, that we will *always* find a way. Here it was, the *people* taking care of the mission, without me worrying about the mission. It was working!

Within an hour, dozens of phone calls, and lots of scribbling on paper, my team came to present me with COAs (Courses of Action). They were able to secure one more UH-60 than previously planned, a bus to transport to the closer spokes, and a reshuffling of personnel to adjust for limited seating in the Blackhawks. These COAs would keep the proposed timeline exactly the same, down to the minute. It was an amazing plan, and I believed it would work. *And it did.*

We met all of our timelines for infil/exfil every day of TAAT and proved even further that the 243d ATCS was ACE capable with very limited resources. We became the premier ATCS of the ANG that week, because our people took a risk and found a way. I'm pretty sure this is *exactly* what General Minihan meant when he charged, *"If you are comfortable in your approach to training, then you are not taking enough risk."*

One of the three ACE capabilities we proved during Triple ACE was the implementation of a homegrown security team whose purpose is to take a handoff from special operators who had already seized an airfield, and then maintain security before the rest of the squadron arrives. This could be a peaceful one-day transition, or it could be a hasty ten-minute handoff followed by immediately controlling airplanes. We exercised both, but the video shows the hasty version. Meaning, we still came in hot, anticipating some potential local resistance, cleared the control tower, and then started controlling the airport while the "special operators" (notional) were cleared off to another location to seize.

Meanwhile, another helicopter full of more controllers and support personnel to set up a landing zone were flown in after our security team had safely integrated and secured the area. Since our governing Air Force Techniques, Tactics, and Procedures (AFTTP) outlines this type of handoff between special forces and ATC, I decided we had probably better know which end of the gun to shoot from, and not look like complete clowns doing it (ask the Army; the Air Force can sometimes look embarrassing when we have our weapons issued to us). So, myself along with two prior Army paratroopers whom I had recruited into my squadron, created the ATC Non-permissive Environment Security Team (NEST).

Nobody gave me permission to do this, because I never asked. I took General Minihan's charge and the Chief of Staff of the Air Force (CSAF), General Charles Q. Brown's "Accelerate Change or Lose" charge, and leveraged my G-series orders to create what I felt was a necessary part of becoming a Combat Airfield Operations Squadron (CAOS).

Creating NEST meant creating a legitimate training program, which we did. Three phases: physical, academic, and practical application in order to graduate and earn the title of NEST Operator. Any career field within ATCS was eligible to apply, which required a resume, letter of recommendation from their direct supervisor, and an interview with the NEST Team Leaders. It took over a year of training, but we eventually graduated ten NEST Operators.

They also received hands-on weapons training (first time our squadron had actually gotten our weapons out of the armory) and tactical helicopter infil/exfil training. NEST is a perfect example of finding a need and taking a risk to fill that need. I feel much better about taking my squadron to war, knowing I can tell families that I have done everything possible to prepare their loved ones to return home safely. NEST also provides an additional "why" to many in my squadron who were wanting more than their regular job. It turned ordinary Airmen into Multi Capable Airmen (MCA). It turned a single bladed knife into a robust multitool.

If someone had asked me to stand up a homegrown security team, I wouldn't have needed a million dollars to do it. All I needed was for those two Generals to charge me with being a damn good leader and fearless commander to ensure my people were ready for war.

I understand that building a homegrown security team from the dust is a pretty beefy example of "making it happen." However, the point is that you can apply these same leadership tactics to the day-to-day, since this is where the foundation is set for bigger things to come.

For example, finding a way to get an Airman advance pay while they are TDY for twenty-eight days, when the rules state that a TDY less than thirty days will not payout until after the TDY is over. If that Airman can't wait thirty-plus days for a paycheck, are you going to find a way to get them paid at the fifteen-day mark like they're used to? There are waivers, exceptions, and loopholes. Find them. Make it happen.

Going back to my pizzas for the squadron example. Did it look like we had enough money in the budget for that many pizzas? Nope, but we found a way to pay for them and also raise the money back for our TAAT dinner.

What about meeting a deadline for annual awards packages to be submitted, so that employees receive their well-deserved time off awards and cash bonuses? Or at the annual awards banquet finding a way to provide a meal and drinks at the formal event where tickets were bought to cover such things, when the caterer fell through the night before?

Or getting a promotion order routed quicker by making some phone calls and hand delivering paperwork to be signed, because due to no fault of their own, the Airman awaiting promotion had the paperwork mishandled up the wickets and it had to get sent back?

What if you were told the projector in the auditorium had just broke, so you'd have to brief off your notes? Could you run and take another projector off the ceiling in a different room and get it hooked up before the meeting started?

These are all personal examples.

Why do we let so many things fail, when a little problem solving, and a high-speed attitude can prevent failure?!

If you were given a million dollars for "mission success," would all of these "small missions" get done? Of course they would. Eventually, you do this enough times and stay on top of your responsibilities, the small things will translate into the larger ones. I guarantee my team wouldn't have had the resiliency and determination to solve the CH-47 nationwide grounding problem the night before our biggest training event in the history of the 243d, if we hadn't spent over two years at that point pushing ourselves to find a way to win the smaller battles. It built our confidence over time, it built precedence, and it built comfort in taking risks.

◆————————————◆

The last "war story" I will use to highlight this leadership tactic, has to do with finding a way to get your people the training they *need*, not the training they are currently authorized to have.

Remember my vision statement? *"Resurrect raw ATC, while modernizing mobile lethality."* In order for my squadron to best posture ourselves to be weaponized by combatant commanders in the next fight, we needed to pay respect to the past. Air traffic control in its rawest form is keeping planes from hitting each other. The very first air traffic controller was a man named Archie League. Back in 1929, Archie would sit on a deck chair at the end of a runway, with an umbrella and flags. He would wait for aircraft to approach Lambert Field in St Louis, Missouri, and then walk out from under his umbrella and wave the different flags of color to assist the pilots with knowing when to land, take-off, or hold (think NASCAR). Can you believe this is how we were controlling airplanes just less than a hundred years ago?

Things evolved quickly, and soon radios, control towers, and radar revolutionized the explosion of air traffic volume that has now overtaken the skies. The story of aviation truly is amazing. So much risk was taken to progress an industry that would change travel, business, and the tactics of war forever.

The world of ATC saw huge gains in capability to detect aircraft, interrogate transponders, and place pertinent information such as altitude, airspeed, squawk identification, and callsigns right onto the controller's radar scope. Hand-offs (transfers) between adjacent control facilities no longer require calling the other controller; you simply enter a letter on the keyboard, click on the radar data block, and wait for the next sector to steady the flashing aircraft with a click of their own. The enormous computers and systems that now assist controllers in ensuring safety of flight in today's congested airspaces is mindboggling.

However, when it comes down to it, all a controller needs is a radio. So, when satellites and radar aren't an option and the only thing you have is a battery-powered handheld radio, you still better know how to control.

So much of the equipment we operate in mobile ATC is heavy, requires multiple aircraft to transport, and takes a long time to set up. However, we now have a military necessity for quicker set up and lighter equipment packages to rapidly jump between locations. Hence why our vision is to resurrect much of the old and raw ways of doing business, while merging these foundational truths of controlling with current technology; the type of technology we can carry on our backs with minimal airfield set up time.

How do we fulfill this vision (promise) without the proper funding or blueprints? I quickly realized I wasn't smart enough to find the solutions on my own. You'll hear me repeat this throughout the book, but I am never the smartest person in the room. I also knew I couldn't hire some big tech firm or aviation giant to crack this nut. Instead, I needed to leverage the human capital right within my own squadron. So I brought my squadron together and issued them a commander's charge: *"Create Capability."*

'Create Capability' was the charge I gave my squadron in April 2021 after having attended our ATC Weapon System Council (the ATC WSC consists of the ten ANG ATCS Commanders). We hold an ATC WSC twice a year, and this specific WSC was especially frustrating. Pretty much, the leadership up at the National Guard Bureau (NGB) didn't have a replacement for our aging equipment, no money to explore other alternatives currently on the market, and no updated blueprint for how to prepare ourselves to become more ACE.

So, I went back to Cheyenne and told my Hawks that we didn't need to wait on DC to tell us what to do and how to do it. I already had General Brown telling me to "accelerate change or lose" and to find ways to meet the demands of the next fight. Therefore, I told my squadron we would create new capabilities with what we had. I wanted them to be innovative, think well outside the box, find the lighter way of doing things, become curious, and I gave them permission to fail at doing all of this. I told them I would provide the top cover, take all the hits for any blowback, and not shoot down any idea that came across my desk. Everything was an option, and the only parameter was to use what we already had. It reminded me of the scene in *Apollo 13* when Mission Commander Gene Kranz charged his engineers with finding a way to conserve oxygen with only the items found onboard the shuttle.

Now would be a perfect time to show a 'Rocky-esque training montage'

115

of my Red-tailed Hawks getting after it. Because they did, and the products they came up with and the loopholes in regulations they found, and the energy behind it all, would've gotten you just as pumped as Rocky running the stairs in Philly. The Hawks delivered!

The biggest wins that came from this charge included a mobile Landing Zone Box (LZB) that my maintenance section developed. It essentially shrunk the capability of our mobile tower (which is on the back of a HUMMWV) and placed everything we needed to operate an airfield into a three-foot hardened box. Boom. They made three of these and we immediately started using them on the dirt landing strip north of our base.

Then someone found a place in Colorado that sewed airport windsocks, and they were able to craft a better landing zone marker; one that wouldn't get blown over when an aircraft landed next to it or took off over it. By the way, a Landing Zone (LZ) is any surface that has been surveyed and marked for landing aircraft. This was becoming our new specialty in ACE ATC. Landing aircraft, anywhere, anytime, on any surface.

Speaking of LZ, the Air Force only has one LZ schoolhouse and since it's such a new skill there are limited seats at the course. So, our two certified LZ controllers found regulations and lesson plans that would allow me as a G-series commander to authorize local area training of our own controllers. This meant they could learn how to control an LZ, but technically couldn't control outside the Wing's training areas.

[Note: As of the publication of this book, the RegAF LZ schoolhouse was shut down, and my unit and the ANG ATCS in NC were selected to stand up the only two LZ schoolhouses in the nation to train Airmen from all components.]

Then we found another loophole in regulations that stated the Group Commander could approve this type of control in their area of jurisdiction, along with Drop Zone Control (controlling the aircraft that are dropping cargo with parachutes). Therefore, when we would attend exercises outside the state of Wyoming, we would get the Group

Commander of that area of jurisdiction and exercise to sign off on our controllers' training and local certifications letter. We were therefore able to "legally" control outside the state.

I already mentioned NEST, but that was also a product of the "Create Capability" charge.

Furthermore, our maintainers located a Deployed Radar Approach Control (DRAPCON) that had been developed as a prototype for the Air Force's mobile radar replacement. By the time it was finished the Air Force had moved onto ACE, and although it was an amazing system, its size and set up time were definitely not ACE. It was used during the aftermath of Hurricane Michael down in Florida to control at Tyndall AFB since the hurricane took out their radar approach control. It had proven to be a very solid system...just not for deploying.

My guys were able to negotiate its acquisition and transportation up to Cheyenne, *for free*. It is a $15M system that we now own and operate out of Cheyenne. This increased our radar scope capacity five hundred percent and has paved the way for increasing our airspace locally. It also made us the first and only fixed-facility RAPCON (Radar Approach Control) in the ANG.

All of this occurred less than one year from issuing the charge to "Create Capability." These weren't our only successes that year. During all of this we also participated in four out-of-state exercises, deployed six controllers overseas in support of OPERATION INHERENT RESOLVE, and deployed five other Hawks stateside to New Jersey in support of the Afghanistan refugee influx after the withdrawal (OPERATION ALLIES WELCOME). The 243d Red-tailed Hawks were singlehandedly resurrecting raw ATC, while modernizing mobile lethality...all while planning our pinnacle Triple ACE training event for that August.

We did all of this without spending a dime over our allocated annual budget.

Leaders, provide your people the permission to be badass, love them when they stumble while trying to do good, and celebrate the hell out of the wins! Empowerment and trust allowed my Hawks to accomplish all of this in less than a year's time, and this was *before* Triple ACE truly solidified our capabilities and put us on the map.

I guarantee any other organization could've done the same things we did if at the beginning you offered them each $1M to create capability. What's inspiring about the Red-tailed Hawks is that I charged them to do this and told them nobody was getting a raise, and we couldn't go over budget. They did it because they already had it inside of themselves, and they just needed a leader to give them empowerment and help lead them to a more intrinsic "why." I love the Red-tailed Hawks, because they have hearts of warriors and always find a way. First there, on the air, anywhere!

Chapter 8

Humility Is Overrated

I have a full sleeve tattooed arm, and a portion of it has the phrase "For the People" in the iconic constitutional font, with a quill subdued behind the words. The reason I replaced "We" with "For" is because I am proud to be part of the less than one percent that currently serves to protect the other ninety-nine. I do it *for* the people. That's what gives me the sense of purpose each day I walk into work, wearing the Air Force uniform and the flag of our country.

To be a servicemember means you are serving *someone*...the *people* the U.S. Constitution protects. This is why part of our oath states that we will defend the Constitution. Although servicemembers are also protected by the Constitution, there needs to be a portion of that population that defends it from the threat of attack, and therefore protects everyone else's freedoms. I absolutely love doing this *for* the people. I love the American people, and I'm not worried about coming off as corny with my tattoos.

When I found out about my Air Advisor deployment in 2023, my wife and I had already bought tickets with our brother- and sister-in-law to go see a popular comedian perform in Fort Collins. I was super bummed that I missed this evening with my wife and in-laws. These are those small things that add up during a military career that make the "thank you for your service" comments from civilians go a long way. Yeah, it sucks missing comedy shows with your wife because you're deployed or TDY. Of course, being deployed during the birth of my child was the hardest thing to miss, but my point is there are plenty of small moments that add up over time. Nevertheless, you're proud to do it. Although my wife was sad that I wasn't there as she sat next to my empty ticketed seat in the 1,180-seat auditorium, she looked around and realized it was completely sold out; my seat was the only one not filled. It made her so proud to know why that seat was empty.

Humility is a character trait that often gets highlighted during leadership seminars or speeches by other leaders. We have placed humility on a sacred pedestal in life, and scorn those who are so outwardly prideful or self-confident. Society always finds ways to hate those that are successful, often by claiming those people are arrogant or cocky.

Haters' gonna hate. Losers will always find reasons to hate winners. Look at any hugely successful person in life, and they have a host of haters. Take Michael Jordan for example. The man was arguably the greatest athlete of all time, who revolutionized basketball and the apparel industry. Yet, when I was a kid growing up in Idaho (where we have zero professional sports and plenty of college teams that are subpar) I was raised to hate the Bulls always winning. Why?! I loved playing and watching basketball, I had never been to Chicago, and I didn't even have an NBA team to cheer for that Michael Jordan kept beating. Yet, it bugged my friends when he would keep winning The Finals.

We are *nurtured* to hate...or should I say, be jealous. Losers hate winners, but guess what? Winners hate losers even more.

It takes a winning attitude to be excellent, even though it can come off as arrogant, conceited, or bragging. And at the end of the day, everyone wants to "be like Mike" whether they admit it or not.

I'm not saying to be selfish or only talk about yourself all the time. In fact, your actions should always speak louder than your words.

However, when it comes time to lead with confidence, you need to be sure of your own abilities. You also can't back down to pressure when you're defending your people or always be worrying about hurting someone's feelings.

I give the description of my "For the People" tattoo and the small example of the empty seat at the comedy show, because I want to emphasize that my pride and how well I do my job, is rooted in my desire to better serve the United States of America. Do you think the American people want a shy and timid military leader? What about an indecisive air traffic controller? Or maybe an Airman that can't confidently speak up when something needs to be said?

When others are so humble that you have to pry their confidence out of them, I think those people are leaving something on the table. It's almost like they are so guarded of their accomplishments and capabilities, because it's more proper to be a quiet badass. Why not let your badassery help other people by letting it be known? One example of this is the two prior Army paratroopers in our unit that helped me create NEST. It took over a year of having these bubbas in my squadron before I finally got to know their true resumes and experiences (the stuff you don't actually put on resumes). These two are super badass, accomplished, and have so much to offer the 243d and NEST! Yet, because of their humility, it took a year for me to tap into that resource. The point of this chapter is to give leaders the permission to be confident and let that confidence and your accomplishments propel the organization you are charged to lead.

◆———————————◆

As described previously, on my first day with my new flight as an OTS Instructor I would first melt their faces with "elevated volume," explain to them how useless they each were, and then proceed to rattle off my resume. I wanted to create credibility upfront so that they would know where my lesson material was coming from (a place of experience, not a textbook) and trust me with their officership training.

I would also walk around campus with a huge wireless speaker and blast my favorite music. *[Note: If you "shuffle" my playlist, you're going to get a wide variety of music, and I really don't care if you make fun of me for it. I own it.]* Most lectures I would start by kicking in the door (I got really good at kicking the handle down and door forward at the same time) with music blasting from the speaker hoisted atop my shoulder. I had fun with it, and I made my presence known. Sure, it pissed off some of the other Instructors in the building (like two), but I still loved those Instructors, and my antics didn't alienate me from being their friend. If anything, it gave us something to laugh about later that day.

I believe some of this mental confidence is rooted in my success physically in sports during high school, but I think it was super amplified by becoming an air traffic controller. In ATC you must be confident. As an ATC Trainer and Instructor, I watched book-smart apprentices fail on-the-job-training (OJT), because they lacked the "badass factor." ATC is the type of job that requires supreme confidence. I grew a lot once I became a rated air traffic controller. I believe this, coupled with my time as an OTS Instructor, prepared me to be confident enough to step into a Squadron Commander position as a Captain.

Remember, the title and purpose of this book is to help others lead above their rank. I hope this chapter's intent comes across correctly; that explaining to you how I gained my confidence, and giving examples of how that confidence was a catalyst for leading, will be helpful in your leadership journey. I'm giving you permission to flip over some tables and do whatever it takes to get your people what they need in order to be successful. General Minihan is a good example of a badass leader who uses his confidence selflessly. You can tell this by the way he speaks and acts boldly, but then only wears unit-level awards on his formal uniform. He epitomizes what it means to be a confident leader that serves the people.

This doesn't mean you have permission to be an ass, arrogantly assume you're always right, or never be a good follower. Again, we have too many of those types of leaders already. Instead, use confidence to serve your people and not get pushed around. Boldly own your style, don't apologize for it, and use it to take care of your people.

•◆————————————◆•

I love being underestimated. In some ways, being a captain helped me get farther as a commander than if I were the appropriate Lieutenant Colonel (O-5) rank. Most people don't expect operational or strategic level performance out of captains; they are still at the tactical level, not commander-level expectations. As a Captain Commander other leaders would confide in me with their insecurities more than higher ranking personnel, because I was "safe." I can't tell you how many other O-5s call me, text me, or visit my office asking for advice that they were afraid to ask their O-6 boss, or didn't want to "look stupid" asking a fellow O-5.

Others see the captain rank and assume they can easily push back on me, thinking I'll be a pushover because they outrank me. These are my favorite.

Early on in my command I had an exchange with a fellow ATCS commander over the phone. He was an O-5 and was hosting the next WSC (Weapon System Council) at his unit. "Due to COVID," he was limiting attendance to just squadron commanders…even though I knew others were coming that weren't commanders. At this time my DO had become my most trusted advisor. He was a fellow warrior and I needed him there with me. As the only other officer in the squadron, I also needed to mentor and groom him to someday take my place. I felt this was a mountain I needed to conquer. One, because it's what was best for my squadron at the time. They needed both of their officers fully represented at the WSC and the 243d was going places; we needed more seats at the table. Two, it was an opportunity for me to demonstrate to my DO that I loved him and would stick by him. Three, being new to the community, I needed to ensure they didn't think they could muscle me around with their rank.

So I emailed this O-5 and professionally informed him that I was planning on bringing my DO anyways.

He immediately called me on the phone.

"Why are you disrespecting me?" he said pointedly. I explained that zero disrespect was intended and outlined the reasons why I needed my DO to attend. I was new to the community, my DO had a lot to offer the WSC, and it was pertinent for the 243d's way forward that he attend. The O-5 responded that he wouldn't allow him in the building if I brought him. I replied if that were the case that I would not be attending.

He backed off. He knew the 243d was gaining traction nationally and that it would look bad not to have all the commanders sitting at "the round table" (since he was the WSC Chairman and hosting the event).

He said I could bring my DO.

I bought him a drink the first night of the WSC and everything was fine. We are good friends to this day and respect each other…regardless of rank.

Fast forward one year later and our WSC was being held in DC, during the same time as the Airfield Operations (AO) Conference where all RegAF leaders for our community were meeting. The new WSC Chairman (different O-5 than the previous story) said we would have to attend the AO Conference virtually, again due to some COVID related pretext regarding space availability. So the ten of us ATCS commanders all flew into DC, sat in a small hotel conference room literally minutes away from where the AO Conference was being held, and listened in virtually to these very important briefings. The conference was scheduled to last a week, and after the first day I was furious that we weren't invited to attend, let alone have a speaking part. ANG ATC was already performing the majority of what RegAF was proposing for future endeavors. We needed to get into that conference, in person!

Luckily our WSC Chairman was able to negotiate for the ten of us to attend the next day.

Still, the week was frustrating as many of us ANG commanders would speak up but didn't have an official voice. On a break I presented our Triple ACE concept (this was before execution) to the O-6 and E-9 in charge. They thought it sounded great, but somewhat brushed me off.

Did I mention I love being underestimated?

Five months later we pulled off Triple ACE. I sent that video to these same individuals. They were blown away. We had a few exchanges back and forth via email. A few days later we had an official invite to the next AO Conference, with a spot on the schedule already blocked off for the 243d ATCS to speak about Triple ACE and our successes. We attended that conference and knocked it out of the park and had Generals coming up afterwards wanting to know more. Since then, our relationship with RegAF has exploded into being stronger than it's ever been. We have made multiple visits to leaders at The Pentagon and were even invited and attended a secret-level conference by the three-star Director of the Air National Guard (DANG). Additionally, that same O-6 from the previous year's conference flew with his team all the way from DC to visit us in Cheyenne and is now a huge catalyst and advocate for getting our squadron converted to a CAOS.

Being a loud and persistent Captain Commander paid off. Humility was not the needed attribute in this particular circumstance. Marketing usually never is humble, and that's what I was doing: marketing my people and their capabilities so that we could obtain the resources needed to improve our readiness for war.

Another example of exuding the confidence needed to take bold actions in an effort to make significant progress, came from my DO, the same one I argued to have attend that first WSC. Rewind to ten months before Triple ACE and the invitation we received by the DANG to attend that classified ACE conference. While I was TDY for two months, the DANG visited Cheyenne. I've explained a little about challenge coins and commander's coins. Most of the time it is commanders and higher-ranking individuals in leadership positions that give out personalized coins. However, there technically aren't any rules governing the "coining" tradition.

One thing I preach to my people is to be authentic, especially around leadership. Even when a General comes walking around, don't act any differently. Be professional, but show them who you really are, and speak freely. I have also instructed my people to ensure that any interaction they do have with a General, is a memorable one...for the General. You never know when you might need to email that General, and you want them to be able to remember you from the thousands of hands they shake each year.

By the way, you don't just email a General...but I have/do.

While I was TDY, my DO was informed a few days in advance that the DANG would be visiting the squadron and that the General wanted to make a quick visit to the 243d (remember, at this time my DO is a First Lieutenant [O-2]). So my DO runs to the Base Exchange (BX) and buys a generic First Lieutenant challenge coin. I don't even know why they sell these. Who is buying First Lieutenant challenge coins? That's not something you put in your shadow box, let alone give someone else. That would be like a flight attendant handing you those plastic kids' wings as they thanked you for flying with them.

However, my DO had a plan to carry out my intent to make his interaction with the General...a memorable one.

The next day my DO gave the General a quick brief on what the 243d is all about, as well as our plans for Triple ACE. From what I was told, he did an amazing job at conveying the vision of ANG ATCSs and

126

specifically the path the 243d was taking toward ACE. At the end of his presentation he proceeded to pull out the First Lieutenant coin, thanked the General for his support of the 243d's mission and future ambitions, and held the coin out in the palm of his hand the same way you do when you're coining someone. Dead serious.

This General stood up with a big grin on his face, reached out, received the coin through the traditional handshake, and they both saluted. My First Lieutenant DO just coined the three-star General of the Air National Guard! Priceless. (I've got pictures from those in attendance to prove it).

Fast forward to the completion of Triple ACE, and I cash in on that moment. I send an email directly to the DANG with the TAAT video link embedded. Again, even if I were an *O-5* squadron commander, you don't just email the DANG directly at The Pentagon...that is, unless your First Lieutenant DO coined him and made the moment memorable.

I'm not going to lie; I was super surprised when this General replied *the same day* and congratulated me on our successes. He wrote that this is exactly what he is expecting of guardsmen, and then put me into contact with his executive to get an official invite to The Pentagon for the classified ACE Lessons Learned conference.

This is the same General that a few months later walked out of his office when I stopped by The Pentagon without an appointment, and with a big smile called me by name, "HH! How are you and the Hawks in Wyoming doing?" Wow. This man is amazing. My SEL and I were walking around to all the top Air Force leadership in The Pentagon, handing out our squadron Christmas card (the one from the Lumberjack Holiday Party with all of us in plaid). This General's executive assistant told us before the DANG walked out that he didn't have much time...yet this man knew my name, opened our card, read the handwritten note on the back, and then took time to go back into his office and coin us with *his* own coin.

This wasn't the only time we found our way to a top-level General.

A few months before Triple ACE, the four-star General of both Army and Air National Guards (Chief of the National Guard Bureau – CNGB) was also visiting Cheyenne. (I know this chapter's timelines are all over the place, but that doesn't matter, each story is standalone and is meant to drive home the fact that humility wasn't the characteristic that got us where we are today.) The CNGB is also a member of the Joint Chiefs of Staff (the Generals who directly advise the President of the United States).

By now the 243d was well known on base in Cheyenne as leading the charge toward CAOS, and the Wing Commander requested that I have a ten- to fifteen-minute presentation ready for the CNGB. I was on the General's schedule that day. Of course I was super prepared and had a room set up in the simulator building that he was scheduled to attend right before my briefing. I was locked and loaded, bro.

The morning of his visit I received word that due to schedule constraints, the General's executive officer had to cut some of the events off his schedule and my fifteen minutes became a casualty. I asked if he was still visiting the simulator and discovered that he indeed would still be in the building (insert evil smile coming across my face). There was no way I was going to let that General out of that building without hearing my briefing. I decided I would creepily follow him and his entourage around in the shadows during the simulator tour, waiting for the right moment to jump out and present myself to him. My hope was to convince him that his most valuable fifteen minutes in Cheyenne would be listening to my brief.

Toward the end of the tour, I weaved my way up to him so that I was literally right off his shoulder, but out of his peripheral vision. My Wing Commander was talking with him in the hallway and saw me sneak up behind the General. If I haven't mentioned it yet, my boss (the

Operations Group Commander) and the Wing Commander are solid supporters and know "my ways." They both give me a long leash and only pull back on it when they see a train coming that I don't. So I could see on my Wing Commander's face that he wasn't surprised that I had snuck up to be literally inches away from the General. Seeing this, my Wing Commander transitioned into, "Sir, I know you have a tight schedule, but I'd like you to meet our Air Traffic Control Squadron Commander, Captain Hochhalter." That opening was all I needed.

The General turned toward me, a little startled at my proximity, and shook my hand. I introduced myself, mentioned something about Oregon where he (and my wife) are both from, and told him I had a five (meant fifteen) minute presentation that would make his entire trip to Cheyenne worth it. He could sense my energy, and luckily, we were steps from the classroom where I had my projector set up ready to brief (it may have helped that I started walking toward the classroom as I kept talking so that he was practically inside the room before he could answer). He agreed (by the way, he was extremely nice and very easy to talk with; an absolute dude).

The General and his entourage entered the room, along with our two-star state Adjutant General...who also knows me and my ways very well, and he just shook his head with a smile. I went right into it and blasted him with a badass, high-speed presentation on who we were and what we were planning to prove during Triple ACE. I threw in what we could use from his level to propel us into the future fight (shameless request). On the last slide, I showed a painting of a cavalry charge (leveraging the quote and analogy that I mentioned earlier in the book) and explained that the 243d was the heavy fast-moving object ready to hit anything and conquer. The painting had an officer on a horse at the front leading the charge with his sword stretched toward the enemy. I then dramatically reached over to my uniform sleeve, ripped our squadron patch off the Velcro, and slammed it on the table in front of him and said, "Sir, thank you for helping lead this charge. *You have our sword.*" You could sense the room hold their breath...and then the General looked at his attaché and nodded; the attaché handed the General a coin. He then looked me straight in the eyes with a steely glare and said enthusiastically, "And you have my full support. Go get 'em, Cap," and proceeded to coin me.

This coin, along with the DANG's, is proudly displayed in our squadron trophy case, because I didn't earn them, the Red-tailed Hawks did. I display these coins as motivation to encourage my troops that they are indeed fulfilling Generals' orders and intent. They are supported by the upper leadership in DC, and they can proceed with ultimate confidence.

Of course, I also emailed this General the Triple ACE video (and later dropped off our squadron Christmas card at his office at The Pentagon). He turned around and mentioned the 243d in his monthly email to all fifty-four Adjutant Generals in the U.S. and its territories. Although he was TDY when we visited DC, his executive assistant showed us his office. He has a credenza with a hutch, and it is completely filled with challenge coins and patches—stacks upon stacks. Guess whose patch was on the very top of the pile? The 243d ATCS patch I had given him off my uniform that day in Cheyenne. Charge!!!

•◆—————————————◆•

Some would say I'm sharing moments of "brown nosing" or "politicking." There is a phrase in the military called "getting face time with leadership" which has a negative connotation associated with it. This usually means the person is only interacting with leadership to get their name out there for future promotion purposes. This chapter will definitely make it come across that I am this way or have an ulterior motive for why I try to interact with upper leadership. That's fine if the reader wants to take this as my objective, but just remember how I started this chapter. My passion is serving the people, both my Hawks and Americans. I would challenge any leader to constantly self-reflect to ensure their intentions are still pure, because these types of successes and interactions with Generals could surely quickly go to my head and make me think I am better than I really am. In reality, I am just the facilitator for the people I serve. I am their spokesperson and I owe it to them to be proactive.

Additionally, I know what it takes to accelerate change, just as General Brown charged us to do. It requires resources that my squadron could benefit from, and how I get those resources has to do with being known (marketing). My Hawks are ultimately in the profession of killing bad guys or being a big enough deterrence to prevent bad guys from attacking us, so that we can enjoy our way of life. My piece in that profession is an important role: controlling airports and airspace in order to gain air superiority. I facilitate the death of those trying to kill Americans. I do this through air traffic. That's the bottom line when it comes to being an arrogant sonofabitch. I want my squadron to successfully play their part in any war and return home, so that all Americans (myself and my family included) can maintain our way of life. That's it, y'all. So, keep this in mind as I quickly review other times I've confidently pushed the envelope, and ask yourself if there are ways you can take similar action to get your people what they need in order to be successful and happy.

◆————————————◆

I had the unique opportunity to be mentored by our Wing Commander when I first assumed command. We both just happened to workout at the base gym at the same time every morning. Although our workouts didn't always cross paths, at least a couple times a week we were both side-by-side on the treadmills. This man is now a General in the state, but at the time was an O-6. I have never met a more personable and down-to-earth officer. He had informally met with my wife and I when we came to visit Cheyenne before I formally applied for the commander position. He was the one that officially signed off on my hire when I was selected by the board. He knew I was coming into a broken squadron, and knew I had an uphill battle as a Captain Commander. He took me under his wing and mentored me during those runs on the treadmill. I will never forget those moments.

He also helped me get a couple meetings with The Adjutant General (TAG) of Wyoming. From there, anytime I saw my TAG, I was professionally persistent on giving him a rapid update on the current status of the 243d. Luckily, I have an amazing TAG as well, and he knows my style. I get a lot of chuckles when I approach top leadership; they know what's coming.

I came to find out that this relationship with my TAG and Wing Commander is unique, as other ATCS commanders at the WSC were surprised when I would mention frequently speaking with these leaders about our future plans. I believe in transparency both up and down the chain of command, and luckily, I have great upper leadership that puts up with my crazy ways and listens when I call, text, or email. I have the support of my state, but it's not because I humbly work in the corner. I enthusiastically let it be known what the Red-tailed Hawks are up to! Fired Up!!!

* * * * * * * *

Eventually I got around to visiting the Cheyenne City Council, Chamber of Commerce, and Military Affairs Committee. I even gave a tour to twenty-five city officials, showed them the entire squadron, our mobile equipment, and the control tower (thanks to our Wing Command Chief having the idea and using his connections to get the invitations out). I had each one of them wanting to join the 243d by the time I was done. They loved it, and now I have been able to make phone calls to these community leaders to gain support in other areas. Since our control tower is a jointly owned and maintained building with the Cheyenne Regional Airport Board and the FAA, these relationships go a long way. It's also nice to be able to text the mayor when needed.

Each year Cheyenne hosts the world's largest outdoor rodeo and western celebration: Cheyenne Frontier Days (most just refer to it as "CFD"), the "Daddy of 'em All!" It's absolutely amazing and has been an annual tradition for 127 years. Five parades occur downtown during the ten days of the rodeo, and the 243d is at every one of those parades: driving our large military trucks down the streets of downtown Cheyenne, while running along doing push-ups with the kids in the crowds as we hand out recruiting merch. It's a blast getting out there with my Hawks, sweating like crazy doing hundreds of push-ups, and building patriotism in young Americans. We blast music and are proud of what we do!

We've also had the Mayor of Cheyenne out for a tour with the local news station to do a story on the 243d ATCS. Public Affairs also has put together numerous videos highlighting the 243d, to include a clip submitting our unit to be virtual participants on "Let's Make A Deal" and a funny five-minute video that my DO and I filmed to be shown during drill while I was TDY at SOS. They also helped put together a video of me speaking to them while I was deployed, followed by past years' photos put to music. This was sent out to everyone in the squadron to remind them of their past accomplishments and let them know that I missed them.

Later, I'll talk more about our mascot and swag, but we've plastered our name all across the nation with stickers and shirts, not to mention locally during our base's biannual Wild West Airshow and annual STEM/Discovery Day. We are the base recruiters' best friend, by frequently setting up our equipment and sending our Hawks to represent at University of Wyoming football games and educators' retreats. I have personally spoken to the Cowboy Challenge academy on numerous occasions and have given dozens of tours to hundreds of people. I even take time to occasionally visit the local high schools and speak with the students in their classrooms about the military. No event is too small.

As a squadron, we collectively participated in sixty-three recruiting events over the course of my first three years as commander. We love getting our people out there and bragging about what we do. Not only has it allowed us to recruit (the right people) like crazy, but it gives our Hawks an opportunity to get out of the office and do something different. They gain pride in being visible to their community while telling their story, and at the same time increase patriotism and support from the community. We have also conducted numerous volunteer operations ranging from weeding and beautifying the Veteran's Hospital grounds, to refurbishing weather equipment and the clubhouse at a local flying club's dirt landing strip east of town.

It's no wonder the 243d won the 2021 TAG Unit of the Year as well as the 2021 and 2022 Cheyenne Trophy. We placed those awards next to the coins we received from the CNGB, DANG, and even the Mayor's coin when we returned the favor and visited his office. Our supporters have grown, and our purpose has been elevated. This is no longer just about a small ATC unit in Cheyenne, Wyoming. It has grown into a nationwide sense of purpose that our people can be proud of, brag about, and find their "why" within.

You can see how sending our TAAT video to anyone and everyone and knocking on doors at The Pentagon to hand out our squadron Christmas card fits right in with our objective to educate the world on the mission and importance of the 243d Red-tailed Hawks. It's about building relationships of unity. Talk with anyone in their seventies or older, and it's the relationships with the people in their lives that they remember and talk about the most. So we strive to make the most of these relationship-building opportunities, knowing this is what will strengthen our American communities and have the most long-term effect on peoples' lives.

Last year I charged the squadron to Educate, Elevate, and Execute. Just because we are servants of the government and not a corporate business, doesn't mean we don't need to market ourselves and achieve brand identity. I don't want to just lace up our cleats for practice every day…I want to play in the actual game! I also don't want my Hawks to run onto that field and not have the training and equipment they need to win. In order to do this, I had to start by educating everyone on what it is the 243d does. Therefore we had to elevate our skill level by participating in higher level combat exercises. Now, we must be given the chance to execute on these skills. Exercises are good, but I want combatant commanders to put us in the game and allow us the honor of executing our mission, real world.

Emailing Generals and paying them visits help. What also helps is getting to speak in DC in front of them, and that was exactly what I was invited to do in December of 2022 during Wyoming's "State of the State" report to NGB. My Wing Commander asked if I would take a moment to show our TAAT video and speak to the DC leadership about our progress with ACE. I was honored to be the only other speaker at that event besides my Wing Commander.

The week before speaking in DC, I was in Georgia at our semi-annual ATC WSC. There I presented our TAAT video and spoke passionately about our future. We had added a few new commanders to the community, and it was exhilarating to get the ten of us ATCS commanders 'Fired Up!' about our conversion to a CAOS, along with getting our people the training and equipment they needed. It was here that the 243d ATCS was officially voted through our charter by the ten commanders to be the first ANG ATCS to lead the way in converting to a CAOS. However, getting to that vote required some blunt words from myself to the other commanders, as well as pointing out that as a captain among lieutenant colonels, I was going to outlast every last one of them in this position. Not only had the 243d proven they were capable and ready for a CAOS conversion but given my "young rank," I would actually get to see this seed sprout and grow, and I wanted to be the first through the door to take any bullets coming our community's way. We'll fail forward and allow the others to come in behind us to do it better, I don't care. I just want to lead the charge.

My SEL and I then spent the next few days driving up from Georgia to DC, stopping at two other ATCSs in South Carolina and North Carolina along the way; meeting with their people and collecting gouge. By the time we finished in DC with the "State of the State" at Joint Base Andrews and our Pentagon visits, we had spent thirteen days TDY, visited three ATCSs, and met more DC leadership than both of our combined thirty-four years of experience had previously afforded. We called it "The CAOS Campaign Tour 2022," and you better believe people are still finding the numerous Hawk stickers we zapped all over their squadrons.

◆————————————◆

As you can see, I think humility is overrated when it comes to leading passionately to get your people what they need and want.

Personally, I think it's okay to let your passion drive communication. I'm not saying I don't occasionally sit on a draft email before sending it when I'm a little hot. However, sometimes (most of the time) straight talk requires some teeth on it to get the point across. We don't have the luxury in the military to sugarcoat our communication all the time. One thing I learned as an OTS Instructor was how to be candid, without getting personal. I never once personally attacked a trainee on any human traits. I may have pointedly explained how subpar their performance was, but I never made it personal. I also found ways to use colorful language appropriately.

Operationally (outside of training environments), most often you will want to take someone aside and not publicly blast them. However, this doesn't mean you can't be direct in a meeting in front of others. I have called someone out in a meeting I held for not writing things down that required action, because they had dropped the ball on a previous deadline. I simply said, "You must be a helluva lot smarter than me in order to remember everything I'm asking you to do, without having to write it down." Does that make things a little awkward during the meeting? Sure. Did I personally attack them or demote them or insult their dog? No, bro. I made a point that I bet everyone else in that room took note of, and then they never showed up to one of my meetings without a pen and paper and didn't bust a deadline again. The key here is not to apologize for bringing the heat. That would be like actually following through as a parent with not giving Timmy that sucker because he wouldn't pick up the ones he spilt, and then later that day apologizing to him for having to be a parent.

That being said, you have to be humble enough to receive the same type of blunt feedback. This is the key. Don't be a hypocrite.

I would like to think I have made more friends than enemies over my career, but I know I have definitely pissed a lot of people off. It's my way of discovering whether or not I can rely on them to handle the heat when the heat is real. If asking you to simply do the job you were hired to do offends you…then I can't trust you in battle.

As a leader, don't be afraid to be passionate and proud of what you and your people have done. Confidence is contagious and I have seen many of my Hawks who were "browbeaten" by previous poor leadership, finally come out of their shell and excel simply because I showed them what pride in their work looks like. In order to do this, you first have to have pride in your own work and have shown you are excellent at what it is *you* do. Leverage that pride and those accomplishments to better your people and organization. Walk with your chest out and head high.

Let it be known that your people can count on you to serve them and their interests. So long as this pride is not self-serving, I don't think it is wrong. Always check yourself, definitely apply humility when it comes to things outside your lane and give the experts the credit they deserve. Like I've said, I'm never the smartest person in the room, but that doesn't mean I can't be the most confident and help guide those smarter than me to reach a collective goal. Be unrepentant of your lethality.

Chapter 9

Get Your Shit in One Sock

When I first assumed command, my boss at the time said something to me that might sound simple and obvious, but in reality, it was hard to implement. "If you only take care of things just the commander is authorized to do, you still won't have enough time in the day." In other words, you will be busy enough with items and situations that only someone with G-series orders can handle, that you shouldn't be doing something that someone else could accomplish. Considering this man had spent a cumulative of eight years on G-series orders as a squadron commander and group commander, I knew he was speaking from experience. He left me to interpret and apply this statement according to my own style of leadership.

I knew for me this meant I had to be even more organized than I already was and be willing to delegate (which is a double-edged sword). I didn't become a commander to be a good manager or an administrative officer. I wanted time in the day to actually *lead* my people. At the root of this chapter is the truth that a good leader is out amongst their people, leading; not behind their computer doing taskers. Freeing up the time needed in order to effectively do this requires a great deal of practical organization and purposeful delegation. I'll come back to delegation, but first I want to address getting your own shit in one sock first (i.e. get organized and don't be a hot mess of scatterbrained due-outs).

This is the part of the book where I provide some practical gouge on what works for me. Nevertheless, do not mistake these types of chapters for "how to be a great manager." Keep in mind how the processes I have found that work are rooted in trying to free myself up to lead, not manage. I know we are all different human beings and have our own mannerisms that play into how we organize our lives. However, I don't buy, "I know my desk is a total mess, but it's how I organize my things and I know exactly where everything is located." This is a load of crap and a dangerous lie to tell yourself. Even if you could dive headfirst into that crumpled pile of papers on your desk and somehow "pull the rabbit we're looking for out of the correct hat," it's not a survivable organizational model for when you actually have shit hit the fan and get swamped. With this type of organizational method you will always be working at max capacity and have no room left for the real emergencies or surges that require you to temporarily increase your bandwidth.

Air traffic control is about staying ahead of the planes, predicting the future, and never straying from your priorities. This requires organization of your radios, radar scope, flight progress strips, and brain. That way, when a surge of unexpected general aviation traffic starts popping up, you aren't already at your maximum mental capacity. Instead, you have room to fit them all in without panicking and can successfully perform the mental gymnastics required to keep everyone safely separated and efficiently routed.

I guarantee your janky-ass office full of randomness all over the place and similar mental organization, cannot handle a surge of unexpected problems to solve. Not to mention the paralyzing image this sends to your people when they walk into your office. I personally would lose a ton of respect for a leader if I walked into their office and it was a total mess, or I had to sit there and watch as they dug through hundreds of emails to locate the item being discussed. You want your people to be confident in your...well...everything. So don't give them reason to doubt. Get your shit in one sock.

[Note: This doesn't mean go out and buy the coolest new "organizational kit" trending around the internet. More often than not, these "tools" just complicate things. Keep it super simple.]

⁌————————————⁍

The best way I was able to explain to my kids what it is like being a commander, was by using a conveyor belt analogy. I would tell them that the "conveyor belt" at work never stopped, it was never turned off at night or on the weekends, and it was never shut down for maintenance. The conveyor belt at work kept a steady pace, 365 days a year, 24 hours a day. It was always set at a standard steady speed, and there were always boxes lined up moving along the way, without much space in between. During the year there would be moments or days however, when the belt speed increased and the gap between the boxes would decrease, meaning many more boxes were being processed than usual. At the end of that conveyor belt was my office. It fed directly into my office and didn't have any other exit. That's where all the boxes eventually ended up.

I would explain that my job was to keep those boxes from falling off the conveyor belt. I also explained that the better organized I was, the further I could walk down that belt and pick the boxes up before they even got all the way down the hallway to my office; essentially giving myself a little break and buying myself time before the next box neared the edge of dropping off the belt…which was always the goal. I needed to be so organized and efficient with my days that I could get ahead enough to run out of the office and make my visits to my people and other places on base to build relations. Once I got back to my office, I wouldn't have any boxes that had fallen on my floor, because I had gotten so far ahead that it afforded me that time outside my office to do the more important parts of my job.

The twist is, I only ever know what's inside about sixty percent of those boxes. These are my scheduled meetings for the day, taskers I'm working on, phone calls I need to make, etc. The other forty percent of those boxes just walk through my door each day and I have to be ready for whatever is inside (as I mentioned in Chapter 5). To be completely honest, maybe this is the air traffic controller in me coming out, but I like the forty percent "mystery boxes" better than the sixty percent "planned boxes." The "mystery boxes" could consist of people just taking advantage of my open-door policy (which could lead to totally throwing off my entire day in order to spend the necessary time to counsel with someone). Or they could be no-notice taskers from the group commander. Or it could be a recruiter stopping by to say hello and talk about a potential recruit. It really is exhilarating to walk through my squadron building doors every morning, only knowing about sixty percent of what that day will hold.

Bottom line, ensure you have an effective and adjustable organizational system which allows you to keep all the boxes from ever falling off the conveyor belt, whether they are planned or not.

So how do I organize my day? I am a very simple man, I'll start with that. I use just a single calendar on my smartphone to input absolutely everything. I live and die by my phone calendar. I have different colors for work meetings, work tasks, appointments (different than meetings), family events, football coaching events, and lastly basic reminders about anything random that I think I'll forget. That's it. All of my life can fit within one of those six colors/categories. Once it's in my phone's calendar, it's off my mind, scheduled, and I *will not* let myself miss a deadline. When I'm in a meeting and set the time/date for the next meeting, I pull my phone out and lock it in right then. If my neighbor is going to be out of town and needs me to pull their garbage out on garbage day, I put a reminder event in my calendar right then. Like everything now, it's backed up in a cloud and I can access my calendar from a computer or restore it if my phone breaks.

I'm not a big fan of Outlook calendar invites for internal squadron related meetings or events. My staff knows to have their own organization methods ready for when I pull my phone out and start spouting off calendar events (which we do in person once a week to stay synced up…more on that during "Chapter 11: Meetings Suck, Don't Suck"). Some of them use their phones as well, some are old school with a paper calendar, and others write it down and then send themselves an Outlook meeting request when they get back to their office. I just keep it in my pocket on my phone at all times.

I do however still use a small leather notepad holder ("padfolio" with the smaller notepad inside) that I carry around with me everywhere during work (along with the water bottle I've had for eleven years; you'll never see me at work without both). I do this because although I use my phone to keep my calendar, I still handwrite all of my notes throughout the day. I don't love the image of typing on my phone during a meeting to take notes. Don't get me wrong, I have my phone out and still use it to look things up pertaining to the meeting, as well as text subject matter experts certain questions when I'm feeling stumped.

So instead I write my notes in my padfolio. At the end of every workday, I transfer any pertinent notes, check off everything I did, rip out the day I just completed, crumple it up, and throw it away. I get a sense of accomplishment from this. I then open my phone and transcribe the next day's events onto that next single half-sheet of notepaper. I start with bullets, formatting my events that have a scheduled time of day associated with them, such as meetings. I then skip a couple rows in case additional meetings come up day-of, and then proceed to make checkboxes for the taskers of that day. Once I finish a meeting or a tasker, I check them off my list. If at the end of the day I wasn't able to check them all off (rare), I transfer them to the next day. This works for me and allows me to add the "unknown forty percent" to the handwritten version of my day for when those moments arise.

Now onto emails. This is where most people get frustrated with trying to keep the bombardment of emails throughout the day from overwhelming them. It's simple. My "Inbox" is only used for taskers I haven't finished yet. Everything else has a named folder off to the side within Outlook that I will file away once completed. Most things I delete, the next most is delegating, then lastly is keeping in my Inbox to complete as soon as possible. If I forward for delegation, I ensure to provide clear expectations in that forwarded email body. I don't just forward and say "see below"...that's super lazy and can lead to confusion on what it is you're asking. Once I forward with the clarifying delegation remarks, I keep that email in my inbox until that person replies that the tasker is now complete. This is my way of following up and not letting someone else fail on my behalf.

I rarely let an email stay in my Inbox more than a day or two. If it's something that requires my signature, it's life expectancy in my Inbox is even shorter. Nothing worse than a process being held up because someone's Inbox looks like my Deleted folder. So I take action on signature requests very quickly. I am protective of my signature integrity, so I'm not flippantly signing things just to get them out of my Inbox. However, I also don't believe in a weekly "signing party" that many other commanders utilize. This is where their administrative assistance files all the items needing their signature into a single folder, then on a designated day of the week they will sit down with their staff and review and sign each item. This is too time consuming and slows processes down even more. One of the many frustrations that come up on government employee surveys is the bottleneck inefficiencies that occurs due to so much paperwork needing to be routed. Just thoroughly review, sign, return, and file. Boom.

The systems I use with my phone, padfolio, and email have allowed me to keep those "boxes" well away from the edge of falling off the "conveyor belt," which frees up more time for me to get out and lead. Again, I was not hired to be an admin officer or a really great manager. I was commissioned to lead. You can't lead from behind a damn desk. Stop making the Marine's joke valid when they ask if you "fly a desk." No dammit! We lead Airmen to achieve air supremacy, so they can bail your ass out of a bad situation by mowing down the enemy you just came into contact with. Keep administrative busy work from getting in the way of your true purpose by being organized and efficient.

As an OTS Instructor I spent as much time as I could with the enlisted Military Training Instructors (MTIs). These are the same professionals I mentioned earlier in the book. I learned so much about how to mold an individual into something excellent from spending time with these feared "trainee hunters." They were always "on the hunt" for something they could make better; a uniform that was out of regulations, a marching movement that wasn't done correctly, or a verbal command given by a trainee that led a flight in the wrong direction. I respected the hell out of these MTIs, and so did all of the trainees (I remember I did as well when I was a trainee at enlisted BMT and then again as an OTS candidate).

I wanted to learn from them, so I attached myself to the MTIs as much as possible. I learned their ways, asked many questions, and practiced my "elevated volume" with them until I got it just right (from the belly, not the throat). Why? Because they always had the full respect of the people they engaged with. Not because they were yelling at them and intimidating. Rather, because they believed in what they were doing and how they were doing it. They understood the meaning of breaking someone down to build them back up. They knew the success of these future military officers depended on their ability to create a disciplined, measured, and resilient officer. So I spent a lot of time learning their ways.

One "trick" I learned from the MTI we all called "Pegleg" (due to his "gansta limp") was how to quickly tighten up an entire flight marching by. He said if you yelled out "cup yo hands!" to a formation marching by, that most likely there were at least a few Airmen not cupping their hands the way they had been taught. The key here was that you weren't yelling at a specific trainee. Therefore, the result would be every member of that sixteen-member flight sharpening not only the cupping of their hands (in fear that they were the one the MTI was yelling at), but also every other aspect of marching that they had been taught.

144

Nobody wanted to be singled out, so *everyone* would "fix themselves" in every way.

I watched him demonstrate as a flight marched by. He told me not to look at their hands to tighten, but instead pay attention to the rest of their stature to improve as the flight would collectively achieve marching perfection. Sure enough, as soon as he bellowed "Cup yo hands!" the entire flight flexed, fixed their arm swings, straightened their backs, stepped in perfect harmony, and, of course...cupped their hands. With one single command the flight suddenly became an extremely crisp marching unit. It was amazing to watch. From then on, I would occasionally walk by a flight and yell "Cup yo hands!" and watch them fix everything. Brilliant.

I started applying this concept to other commands. In the mornings when we'd wake up the trainees at 0430, a rush of hundreds of officer trainees would come pouring down the stairwells and hallways, all to exit out of the same door to form up for marching to their first training event of the day. They were required to "square their corners" as they marched hurriedly through the tile floored hallways toward the exit. I would stand near the exit at a random corner and yell out "square yo corners!" Same principle, same results. Brilliant.

The reason I tell this story is so that you understand what I mean when I say to "square your corners." I have given this charge to my squadron in the past and told the same story so that it can be used in any context. It's another tool for my people to use in order to hold each other accountable.

I placed it in this chapter because I feel it's appropriate when talking about organization and keeping yourself "squared away" when it comes to how you execute your daily schedule. Is there something in your way of operating/organizing versus mine that could use some "tightening?" Are you truly efficient with your time? Have you tried different organizational and scheduling methods to ensure you never miss a deadline and are never late to a meeting? In what ways can you "fix yourself" after considering maybe you're part of the equation for inefficiencies within your organization? A good assessment to know if your methods are working is to take an honest account of times when you've missed a deadline or failed to get someone something they needed or caused an unnecessary delay. Afterall, leadership is service, and if your people aren't getting the service they deserve because you keep failing to remember things or provide clear instructions, then maybe you need to reassess how you organize yourself.

Listen. You're a damn leader. Leaders don't show up to meetings late, leaders don't miss deadlines, and leaders don't have a pile of ass on their desk that "needs attention." No, bro, leaders square their corners and cup their hands. We don't let complacency sneak in and sabotage our squadrons or companies. Because remember?... *It's your fault* if your organization sucks. Get your shit in one sock.

◆———————————◆

Being very well organized not only helps you execute your daily schedule more efficiently, but it also amplifies your leadership presence, meetings, and communication. Another item that always shows up on government employee surveys is poor communication. This is a difficult target to hit, since this could mean so many different things. However, for my squadron specifically it meant finding new ways to keep lines of communication open and transparent with our DSGs. They don't typically have government laptops at home or some other way of accessing information that requires a Common Access Card (CAC).

The approach I took was trying to find a platform that allowed instant sharing of unclassified information, accessible on every squadron member's smartphone. Since we always have our phones and are constantly checking them, I wanted to make it just as easy to communicate back and forth with my people as text messaging. So, I put my team to the task, and they found a free application that allowed us to do just that. We didn't have to reinvent the wheel, we didn't have to ask our people to purchase anything, and most importantly we didn't compromise operational security.

Once all members had the app downloaded, we started to upload documents, unclassified information, and question forums. Staff meeting notes and presentation slides are uploaded weekly so that even part-time employees stay up-to-date. We have someone assigned to be the app manager to keep things updated and current. The first few months we had to work through some unknowns, but it is now a well-functioning communication platform which is organized and efficient. This has solved the vast majority of our communication gaps and has also shifted the responsibility of "being in the know" to our people. This is because we push everything we can safely share to the app so that there is no excuse for "not knowing" or "not getting that email." No bro, it's on the app. You just didn't look. So it increased communication as well as personal accountability. Organization was and remains key to this app's success.

Much of organization has to do with prioritizing. I like using the analogy of a big glass jar, with big rocks and sand needing to fit inside the jar. The glass jar represents your day. If you pour all the sand in first, then there is no way you'd fit all the rocks inside. Instead, you must carefully place the big rocks in first, and then pour the sand in last, while gently shaking the jar to allow the sand to settle in between the big rocks.

Your day should be just like this. That is why I transcribe my next day's schedule from my phone's calendar to that single half sheet padfolio. My phone mostly captures the "big rocks," but a lot of the "sand" are things I'm working on or needing to accomplish the next day that don't have a scheduled time. So writing at the top the "big rocks" in bullet form with times of the day associated, and then taskers (my "sand") below with check boxes, allows me to visually see everything I need to fit within the "glass jar" before I start my day. It's also similar to packing a moving truck (which after fifteen moves in eighteen years of marriage, I've gotten unwillingly good at...and I now break out in traumatic sweats whenever I see a moving truck driving down the road). You've got to bring the majority of the items needing to be packed out to the driveway to see what you've got to fit in. Then you start with the big items and move your way down to the smaller objects to fill in the gaps.

What writing your schedule out visually also does for you is it allows you to be flexible. Everything I've written up to this point in this chapter sounds pretty rigid, because that's what organization is; it's structured, it's an effective system. However, because of how well organized the "known sixty percent" of my day is, it allows me to be flexible with the "unknown forty percent" that walks through that door or gets shot my way last minute. So as much as my ways of organization are very systematic, the reason for this is so that I don't have to be rigid with the extra "sand" that always ends up needing to fit into the jar by the time the day is over (which mostly includes "people time"). So please, don't be so focused on checking off your boxes that it prevents you from doing the most important part of your job: leading by taking care of your people.

This brings me to delegation. The reason I call this a "double-edged sword" is because delegation can get you into trouble quicker than it can help you if done incorrectly. This is due to leaders either shoveling their problems onto subordinates, or properly delegating but not effectively following up. In order to utilize the very necessary tool of delegation you must first ensure you're "sending it down the right highway." My squadron is organized to where I have three direct reports to the commander for operational execution. I have a military Director of Operations (my number two in command), a military Senior Enlisted Leader (the most senior enlisted member of the squadron and my most trusted advisor), and a DoD civilian Air Traffic Manager (whose primary responsibility is to run the Cheyenne Airport ATC system). All three report directly to me as their supervisor.

Although each of them has overlapping "lanes," they still have very distinct roles.

When I first took command, I drew out a Venn diagram with three circles. I wrote inside the circles to show just exactly what each of them owned versus what they shared. This helped a lot, but it's a continuously evolving process. I have sometimes wrongfully delegated something to one of these three, just to have it come back that it should've been in someone else's lane. Most of the time it wasn't the person I delegated to telling me it wasn't their job. This has never been the case actually, because I have studs that work in the 243d and they understand that this common line ("it's not my job") which many other government employees pull, will get me 'Fired Up!' quicker than anything…and not the positive version of 'Fired Up!' It was actually the person whose lane it should've been delegated to, asking for it themselves, which was impressive—to be asking for more work.

However, the point I'm making is that I sometimes mess up and don't give it to the correct person. Obviously, over time and after working together for so long, the four of us have gotten very efficient with proper delegation lanes. Open communication is key, which requires you to be an approachable leader so that your people aren't afraid to tell when you've possibly made a mistake.

The second part of delegation is the follow-up. Once you've determined who should be responsible for the tasker you're delegating, you need to provide three things: *background, intent, and timeline* (or "BIT" which is easy to remember, because usually you delegate when you've "bit off more than you can chew"…yeah, I made all this up, but it works for me).

Nobody wants to have a bag of shit placed on their desk and then you just walk away. At least explain why the bag of shit is on their desk, where it came from, and any amplifying information to assist in getting it removed from their desk as soon as possible. Along with the *background*, you must provide your *intent*, which means explaining your desired end state or at a minimum the manner in which you want this handled. We all know there are numerous ways to accomplish the same task. Provide the wide or narrow path you desire, or else don't complain about how it ends up getting done.

Lastly, set a *timeline* for periodic updates if it's multi-phase and going to take a while, or a hard deadline for completion if it's a short-term tasker. Once you've properly picked the correct lane and done your three parts to delegate appropriately ("BIT"), your job doesn't end there. You must either put it in your phone, tag an email in your inbox, or write on your paper planner when to *follow up* with this person. I don't necessarily write down every tasker I delegate, this would nearly defeat one of the purposes of delegation, which is to get the subject matter experts on the tasker (this means empower your people, it doesn't mean be a micromanager).

The other purpose of delegation is to free up your time for things like...I don't know, maybe leading! So be careful not to go too heavy on the follow up. Trust your people to get the job done. I've gotten to the point with my people since most of us have been working together for years now, that I don't need to make a follow up note. I know once I send it down the proper lane and provide the three items they deserve from me, that they are going to get right on it, and that they will report back to me once it's done. Hell yeah! Delegation works! However, if you skip even one step of the process I just explained, it's your fault when it doesn't get done. I mean...it's your fault either way, so do your part and be a clear communicator!

◆━━━━━━━━━━━━━◆◆

One last experience I'll share to drive home the points of this chapter relates to the time I attended and graduated SWAT Basic Tactical Operator school. While at OTS as an Instructor, a very rare and random opportunity came up for a few of us Air Force officers to attend Alabama's SWAT school. Although I had zero justification to attend, other than I like testing myself and learning new things about my capabilities, I was selected along with two Security Forces officers to attend. Keep in mind at this point I had graduated Air Force Basic Military Training, Officer Training School, and was currently teaching at OTS. So, I had been yelled at and done the yelling, many times. However, this seven-day SWAT course was a whole other level.

Over the span of the seven days of training we only logged a total of fifteen hours of sleep. We were fed less than well, teargassed, rucked, put on our faces more times than I can count, conducted tactical helicopter infils/exfils, put through physical fitness testing, weapons training, shooting qualification, tactical first aid on raw pig meat, rapid clearing of school buses, conducted night vision drug interdiction (that's where I got an arm full of poison oak, low-crawling through the dense Bama woods), and hours, upon hours, *upon hours* of close quarter battle (CQB) training; clearing schools, courthouses, and trailers.

The Instructor team was topnotch and kicked our asses around, nonstop. It was awesome. We started with thirty candidates and graduated seventeen by the end of the week. It was a proud moment to be awarded the SWAT badge and Tactical Operator certificate, not to mention humbling to spend a week with law enforcers from all different agencies (city police, state troopers, U.S. Marshals, etc.). I gained a great deal of respect for what these men and women do every day to keep our local communities safe. It was an honor getting my face kicked in alongside these bubbas.

You're probably wondering what SWAT training has to do with getting your shit in one sock and being organized? First, when we were clearing large buildings (such as a high school) with thirty of us, we had to be extremely organized. Who was going to breech, who enters the room first, who covered the entry while the room is being cleared, etc. With so many of us quickly moving down hallways and in/out of classrooms, we each had to know every part of the job since we would literally run through the hallways and constantly be switching roles depending on who was in which order in the stack. We were taught to move through the building like water; constantly moving forward with purpose, flowing through hallways and rooms to check every corner and under every desk, and not stopping to think twice or be stagnant.

Second, we were told to always "Find Work." What this meant was if we entered a room with three SWAT operators, once we cleared a corner or area, don't just stand there pointing your weapon at an empty cleared corner! Find work! See what needs to be done next. Move onto the next room or cover down on the next stack entering a large office at the back of the room. Don't just stand there patting yourself on the back for having cleared a corner and then hold it down. This simple mentality to always "Find Work" did wonders at moving a thirty-member SWAT team through a huge high school, efficiently and very quickly.

I challenge each of you, that if you get so good at being organized and having your day running efficiently, FIND WORK! If you are sitting behind your desk, staring at an empty inbox, a checklist that's completed, and nobody walking through your door...congratulations! You've officially gotten to the most important part of your day: time to lead! Get your ass moving and put your people first and always. Conversely, if you're discovering you aren't getting it all done to the point you find time to lead, then you need to make some serious adjustments.

First, look to your top-level managers and ensure you have provided them the proper training needed to do their jobs extremely well. If you have, kick them into gear and provide clearer expectations and measurable performance standards. Then ensure you're actually doing all three steps to delegate effectively (BIT: background, intent, and timeline). Hold your upper- and mid-level managers accountable (remember "The Three Principles of Unified Power"? If not, reread "Chapter 6: You're Either Coaching It, or Allowing It").

I would also suggest walking around your workplace and speaking with the lowest ranking employees and getting their views. They know the most about why things aren't running efficiently, but nobody ever asks their opinion. All of these tactics can help free up your time as the leader, which will essentially allow you to take better care of your people, which will generate time for inventive discussions, which will ultimately advance your organization steadfastly forward.

Remember, only do the things only the leader can do (which aside from the signature authority paperwork is actually finding better ways to lead your people). Every day you walk into work, that is the goal. Expertly manipulate the big rocks and sand so that you effectively get outside your office and start leading. Find Work. Find your people. You're a leader, not a manager...but you must manage your time, to be an effective leader.

Chapter 10

Don't Look Like a Bag of Ass

Have I offended you yet? If I have, thank you for still reading…your skin is getting thicker the more you read. If I haven't offended you yet, I'm going to try my best in this chapter. The reason I say this is because this chapter concentrates on your appearance, and if there is one thing that most people always take offense to, it's when you comment negatively on their appearance. Just a warning. Also, it's short, so you can easily reread it when you start slacking off on your dress and appearance.

One of my favorite old military sayings is, "When does a good Airman need a haircut? …never." This means you always get a haircut before it's even close to pushing the limits on military grooming standards. The same needs to apply to every aspect of your appearance, because guess what? It influences your performance. *Appearance affects performance.*

I am not just speaking to military members. This book is intended for all wanting to become excellent leaders *now*, even before your rank and position formally expect that of you. I don't care if it's the military or a call center, you always need to look "high-and-tight." The best is when you don't even have a dress and grooming standard; it provides even more opportunity to distinguish yourself as a professional.

The summer I spent attending the FAA Academy in Oklahoma, I wore civilian clothes to work and there were no dress and appearance standards. Students wore anything from shorts, a t-shirt, and baseball cap, to what appeared to be their pajamas they woke up in that morning. Most of them looked like a bag of ass. They were my friends and I got along with them just fine, but they didn't look professional. They didn't have to. The FAA didn't care, since their customers (the pilots) never saw them. They only heard their voices over the radio.

Opposite to their slacker apparel, I wore chinos, a tucked in button-up long sleeve shirt, and leather shoes every day. I ironed every outfit the night before work. You better believe I stood out. I also graduated number one from the academy's most intense course and set a record doing it. Go ahead and enjoy the comfort of your pajamas while I show you how to control airplanes. I'm a big believer in your appearance having an effect (positively or negatively) on your performance. I also believe in dressing for the position you want, and not waiting until you get that promotion. Sound familiar? "Leading above your rank…punching above your weight…just because you're a lieutenant doesn't mean you need to act like one…" It's all connected.

For those that do have dress and appearance standards, find ways to elevate the norm. Although there are certain restrictions to altering a military uniform, there are a lot of ways you *can* improve its appearance.

I have always ironed my uniform and ensured it was well-fitting. However, becoming an OTS Instructor and spending so much time with the MTIs who were experts at making their uniforms even more high-and-tight, I found my way to another level of professionalism.

I went to the local tailor and made adjustments. I had them taper my pants and jacket (called a "blouse"), removed certain buttons and replaced them with Velcro for a more streamlined look, sewed a couple pockets down, clipped off a few zipper pulls, placed a hat former in my patrol cap, pre-bloused my pant legs, and sewed down all my name tapes and badges. I took it back to the same seamstress three times before I was satisfied (bless her heart). She was a sweetheart (kind of) and eventually called my combination of alterations "The Hochhalter Special" (which she named tongue-in-cheek). Each OTS class, I would have trainees approach me asking where I got my uniform tailored. I would tell them where, which seamstress, and to get "The Hochhalter Special."

I want to make something clear, I'm not trying to win a beauty pageant. I'm taking pride in everything I do. One of the Air Force Core Values is "Excellence in all we do." If I want to be excellent, I need to dress excellently. This also means taking good care of your uniform. I get it, the camouflage uniforms are now made to be "wash and wear." However, they aren't meant to be "wash…and then dump on your floor and never iron or even put on a hanger." Don't do that. Have some pride in the cloth of your country and don't look like a bag of ass.

Would it hurt to keep a hanger in your car and take your blouse off to hang in your back seat to-and-from work, so it's not wrinkled by your seatbelt and avoids spills from your morning coffee? How about you take an iron to at least the collar so it's not flicking you in the cheek every time you turn your head? Or better yet, wear the proper socks and not whatever color of nasty tube socks you blindly pulled out of your dresser that morning? Gotta love seeing your hairy legs when your pants come up past your boots to reveal a too short and wrong colored sock. Nice. Looks fantastic.

Also, please stop wearing your military undershirt out in public after work. It's obvious enough that you just got off work because we can see your dimpled boot and sock lines on your legs while you're wearing baggy basketball shorts and house slippers to the BX. Can you at least change your damn shirt? Yes, I'm talking to you, dorm rats living on base…you know who you are.

You are always on parade.

Hold up civilians, you're not off the hook, get your butt back here. Just because your organization believes in "free thought, imagination, and unicorns" doesn't mean you should dig out whatever outdated wrinkled shirt you can find from the bottom of your dirty hamper, smell it, shrug, and then pull it over your greasy-haired head. How about you shop for some clothes that fit, iron them, and shampoo your hair? Combing your hair would be extra special. Stop succumbing to the new age thought of "If it tastes good chew it, if it feels good do it." Nobody wants to smell what you just "chewed," and you probably just caught a disease "doing" whatever it is you just did. So snag yourself some breath mints at the gas station and stop bouncing through life expecting things to be handed to you. Work your ass off for it, look the damn part, and take pride.

This also goes for how you organize and decorate your office, as well as how you take care of your building...including the parking lot.

Even when I just had a cubicle, I took pride in making it professional. My cubicle at OTS was nicknamed "The Den" because it was decorated like a sophisticated office with "many leather-bound books..." It helps that my wife is a certified interior decorator and has a natural touch for this type of thing. She decorated my current Commander's Office so well, that it certainly demands respect and reflects the Office of Commander. It makes a difference.

Also, my squadron building. I have renovated things quite a bit. Paint, camo netting, framed mission and vision statements, better chairs, large vinyl decals on the walls, blinds opened, windows washed, floors always vacuumed, garbage never overflowing, etc. I had my parking lot restriped the year I took command and bought new reserved parking signs for the four of us that have them (the old ones had vinyl lettering peeling off and were super sunburnt). Anytime I walk down the hallway or out in the parking lot and notice even a small piece of trash, I pick it up and throw it away.

I have pride in my squadron, and I want my squadron to be the best ATCS/CAOS in the entire Air Force. Will I become the best squadron because my walls are painted, and my uniform looks great? Not by those things alone, no. However, those little things coupled with the other tactics I write about in this book have led us to become just that: the number one ATCS in the ANG, voted first to convert to a CAOS, and one of only two units selected nationwide to stand up a landing zone schoolhouse. So yeah, it makes a difference. It's a culmination of tactics. Whenever my crew checks into a hotel for an ATC Weapons System Council with the other nine ANG ATCSs, you better believe the others notice the business casual travel attire. Damn straight.

———————◆————————

The last part of this short chapter deals with your fitness. Again, this doesn't just apply to the military readers. Civilian leaders could do well to take care of their health. If you're doing it right, leadership is going to be physically demanding, and being active and eating right is going to make you a better leader. However, specific to the military, we have an actual *obligation* to the American people to be physically fit. You must pass your annual PT test in order to maintain membership in the armed forces. The Air Force PT test standards are incredibly generous. You do not have to be anywhere close to an athlete or gym rat to pass your PT test. So, when you score below a seventy-five percent, it's really bad, bro. Like, really.

Leaders, you shouldn't even be close to the seventy-five percent minimum. All officers should be above a ninety percent, and even that's not asking much. As of the writing of this book, my 1.5-mile time is 8:42 and I max out push-ups and sit-ups as a thirty-seven-year-old. I am the fastest in my squadron and score one hundred percent every time. I am also one of only two officers. I am the commander. Don't you think that's the way it should be? If you had to guess how many hours I work out a week to achieve that type of time and score, what would you say? Five hours, bro. I work out exactly one hour Monday through Friday, and I take the weekends off. I run for an hour M, W, F, and do mostly bodyweight muscular training T, and Th. I also eat pretty clean. It doesn't take much.

Can you guess by now what I'm going to say? *It takes consistency and follow-through.* That should be a reoccurring theme in this book by now. That's a little sneak peek to Chapter 17: "No Secret to Losing Weight." ...and here's another hint, I'm not going to be talking about actually losing weight, y'all.

Earn those "Thank you for your service" remarks you get from civilians when you run into the grocery store for milk on your way home in uniform. Don't make them say it out of obligation, because they see some fat ass testing the thread strength of their uniform, or they can't read your Velcro name tape because it looks like a damn strip of bacon. Earn that thank you. Be the less than one percent the other ninety-nine wants to see charging the hill against the bad guys...not rolling down the hill...like a bag of ass. Enough said. Don't take offense, just take care of business.

Chapter 11

Meetings Suck, Don't Suck

Be brief, be funny. If every meeting you attended was brief, and had some humor sprinkled throughout, wouldn't it be easier to *not* want to slam your head on the table halfway through? I think it's a pretty simple and proven tactic—be brief, be funny. Just like teaching is more effective when you "edutain."

The point of meetings is to generate action. As I explained in Chapter 6, one of my favorite quotes from my commander's course was "What you *do* out there, matters more than what you *say* in here." That is my mantra for every meeting I hold. The purpose of this saying is to remind those in attendance that words are cheap and that nothing we say in a meeting is worth a damn if we don't *do* something about what we just talked about. That is why I believe in an "around the room" summary at the end of every meeting. This is so that every person has an opportunity for any "saved rounds" as we call them in the military (something you wanted to say during the meeting, but didn't get the chance to say), as well as to state what actions (or we call them "due-outs") you took from the meeting. I'm getting a little ahead of myself, but I wanted to explain the intent behind the quote.

First, I want to go over meeting structure. Don't worry, I'm not going to be a hypocrite in this chapter and bore you with things you probably already know. I just want to provide a very proven method for holding successful meetings. I also want to continue to point out that even if you are not the official leader or didn't call the meeting, you can still help guide the meeting in the right direction.

Meeting structure begins with no structure. When I was in high school, I took German for a couple years (don't remember a damn German phrase now). As soon as the bell rang at the beginning of class, the teacher immediately started lecturing and teaching, picking up right where she left off the day before. It was horrible. No small talk, no "how's it going class?" no lead-in to what we were about to cover that day. It was like pausing partway through a boring foreign movie with English subtitles, and then pushing play again the next day from the exact spot you left off. She was also dry and boring (hopefully she never reads this book or doesn't remember me).

Conversely, my history teacher would start class by finding something funny to say about something a student said or ask what anyone heard was going on in the news, or if anyone went to the football game last week. It put everyone at ease and never took more than five minutes. It was completely unrelated to the topic of history, but it made me look forward to class and got my mind engaged. I remember a helluva lot more historical facts than I do German phrases.

So, this is what I try to do when I start any meeting I am in charge of. Just some random bullshit. Impromptu. I don't go into the meeting with any idea as to how I'm going to open it, I just feel it out and find something arbitrary. It could be as simple as asking a few people what their favorite cereal is (amazing what you can find out about a person with that question). I do things like this for a few minutes, which also allows me to get a feel for the room. Then at a natural break from the small talk, I start in on the "why" behind the meeting.

Outlining the "why" of a meeting upfront, along with no more than three objectives, provides you as the leader of the meeting an opportunity to establish expectations from the start. Undoubtedly, the meeting will stray off topic. However, since you made it clear from the beginning what your objectives were for that meeting, you can tactfully redirect any "bogey" comments without it sounding too harsh. In my opinion, this is the number one reason meetings go so long: the meeting takes a turn off topic and the leader doesn't quickly negotiate a redirect. As the meeting leader, you must bring it back to topic immediately. There may be some valid comments being made and you'll want to dive down that rabbit hole. Don't do it! Jot down a quick note if it's valid, and either categorize it into another meeting you already hold regularly or continue the conversation "offline" after the meeting.

I am a firm believer in meetings being less than an hour long, with forty-five minutes being my ideal target. If whatever it is you're talking about requires more than an hour, then you should attempt to cover less objectives/topics and schedule separate meetings. I would rather have two separate one-hour meetings a couple days apart that concentrate on separate topics, than a single three-hour meeting where you try to cover too many unrelated topics. You are going to lose your people and the level of productivity if you go over an hour, and you'll inevitably waste time spinning your wheels and taking breaks if you cram too much into a meeting. Stop it! Be better at planning the objectives of your meetings, stay on topic, start and end on time! Period.

Mine is the "rule of three." I will never hold a meeting that has more than three topics/objectives. Since I max out at an hour (and I drink a ton of water throughout the day, so my bladder can't go longer anyways), that only provides about eighteen minutes per topic, if you conduct a five-minute ice breaker. You can see how difficult it could be to keep the meeting to just an hour, even if you only had three objectives/topics. So don't hold a meeting with more than three!

Once you've softened the crowd with three to five minutes of small talk and outlined the objectives/topics of the meeting, now it's time to play "auctioneer." I've stated it before, but I am never the smartest person in the room. Mark Twain once wrote, "It ain't what you don't know that gets you into trouble. It's what you know for sure that just ain't so." As leaders with authority, we have to be careful not to fall into the pit of illusory superiority, or superiority bias. That's why when I hold meetings, I want smarter people than me in the room to advise and problem solve. Otherwise, why wouldn't I just sit in my office and do it all myself? And there's my point. You can't. That's why you hold meetings, so that you can collectively be smarter than the individual. Keep that in mind when you're choosing who to invite to the meeting, and also when you feel the urge to start shooting everyone's ideas down. You need them, so don't immediately murder their ideas without impunity, using the energy of a blackhole sucking their thoughts from their very souls and crushing them while they watch in horror.

A good meeting leader or "auctioneer" moves from person-to-person, seeing what everyone has to offer. Don't let the "me monsters" steal the show either. Sometimes it's the quiet ones in the corner that have the most to offer, they're just too timid to speak up. Pull it out of them by bouncing around the room like an auctioneer and ask open-ended questions to generate better discussion and solutions.

One tactic I use to ensure I do this is ending the meeting by quickly going around the room and pointing to each person one at a time. "(Name), anything else?" Be careful, this could quickly make you go over time if you haven't managed the meeting properly up to this point. However, so long as you've orchestrated the meeting well and moved around to everyone, this shouldn't take more than five minutes. It is also expected that at this moment when you point to each person, they report on what their "due-outs" or "actionables" are. If they don't, you should've been writing them down yourself (since you're going to follow-up with them, right?), and it's at this moment in the meeting that you can remind them.

This is also a great time to publicly give credit to the "heavy lifters." As a leader, you're often skillfully orchestrating the actual work, not "in the trenches" doing most of the taskers. If you're delegating correctly and sticking with those things nobody else but you are authorized to do (as outlined in Chapter 9), then stay out of the lanes of those who own a specific process and stop micromanaging! Then, take time during meetings to recognize those doing the actual nitty-gritty work. Pause to think about how long it must've taken that person to build that spreadsheet, or track down the correct information, or visit that office on base and put up with the crusty old retiree who is hoarding information just to guard what they think is an important job, but deep down inside they know they are irrelevant. You're lucky you don't have to run somethings to ground like you used to before you got this leadership position, so don't forget where you came from, and give credit where it is due.

Also, don't act like you're still as good as you once were at whatever it is your frontline employees do. I get it, I used to be a badass line controller, but guess what? ATC is a perishable skill, and the actual controlling of airplanes is no longer my primary duty as a commander, so I'm not going to walk into the control tower and take over controlling the airspace during peak traffic hours. Instead, I'm going to get my few hours of monthly proficiency time during the part of the day when "Ol' Boy Crop-duster" and "Kit Plane McGee" are the only ones in the sky. I'm not going to insult those on the frontlines by talking about my glory days, and then require assistance when I get more than three planes. It's all right to admit to your people that you don't have the answers nor the skills to do this leadership thing alone. So, during a meeting be sure to recognize those that still do the heavy lifting, and don't throw down some old-school technique that is already outdated.

Nevertheless, you are the meeting leader so don't abdicate that position and allow someone else to commandeer your meeting. This is a leadership book, not a managerial book. Although these are practical management tactics, the intent is not to waste time away from the actions needed in order to take care of the people. Mission accomplishment depends on effectively and efficiently getting your people to operate at a higher level. This isn't done by holding two-hour meetings about topics unrelated to the people and the mission. As soon as you allow a meeting to get off topic, you are wasting time, and the enemy appreciates that. If your enemy is another business that you're in competition with, you just gave them precious additional hours toward their efforts to destroy you. Don't waste time and always stay on topic.

Just like everything else I've written in this book, consistency is key. Choose your regularly scheduled meetings carefully, so as not to overbook your people. I hold just two weekly recurring meetings within my squadron. The first is a Monday morning CUB (Commander's Update Brief) where my command staff and I review the calendar for the week, along with focus areas and priorities. It allows me a formal touching point with my command staff at least once a week. Since we are all busy and have other meetings outside the squadron throughout the week, it's sometimes difficult to get us all in the same building at the same time. Therefore, this guarantees at least once a week we will all be together in the same room.

This doesn't mean I don't occasionally call a "hallway huddle" when I need five minutes of everyone's time randomly during the day. I simply yell out, "hallway huddle!" and everyone exits their office and stands in the command staff hallway (I may or may not sometimes do this just to throw random office objects at them such as dry erase markers, in an attempt to break up the monotony of the day). I also hold a standing meeting with my staff in my office ten minutes before every commander's call (my monthly address to the entire squadron). This way we are all on the same page and a united front concerning how the squadron-all meeting is going to run (since they are different each time).

The second weekly meeting is our Squadron Staff Meeting. This consists of all my upper- and mid-level managers as mandatory attendees, however anyone and everyone are invited to attend. This is held on Tuesday afternoons. We have a slide deck we keep updated (I am very picky with the number of slides I allow), and we have a very proven system for getting through a lot of information in under an hour.

And that's it, dude. I run a ninety-four-member squadron with just two *recurring* squadron-level meetings a week, totaling a max of two hours. Yes, there are other once-a-month meetings, and drill weekends are super busy with additional meetings, but the majority of the collaboration occurs during these two events. I also like to hold appointments for very specific purposes; many times these are held in my office, depending on the size (we have a conference room at the beginning of the Command Staff hallway for the other larger meetings). These appointments are very surgical, and even shorter. I prefer multiple twenty-minute-or-less meetings scattered throughout the week, over a few long meetings where a wide array of topics are covered. Get in, develop an action plan, get out. We should be spending more time *doing* than talking. We are a squadron of action.

[Note: Although consistency is key, do not hesitate to cancel a regularly scheduled meeting! If no new information is going to be passed and nothing significant requires convening, then give some time back to your people and cancel the meeting. This is why I require all slides for the Squadron Staff Meeting to be updated by noon the day before. I have those who "own" a slide put the "Current As Of" (CAO) date at the top of the slide. By close of business the day before, I can quickly click through the slides. If I notice most CAO dates haven't changed (meaning the information has stayed the same since the last time we met), that's a pretty good indication that I probably don't need to hold the meeting that week. If you find yourself canceling often, reconsider the purpose of the meeting and possibly combine it with another one, so long as it won't exceed the three objective rule or put that other meeting over an hour.]

You need to know when you're oversaturating your people with meetings, slide presentations, emails, texts, and even office visits (most leaders definitely could increase their casual office visits but be careful not to slow down productivity when someone is giving you the non-verbal social cue of, "fun fact: I don't care, boss, and I need to get back to work"). Sometimes I have to back off the texting, calling, and shop talk.

To help combat this, I've consciously tried to create a physical point for my brain to transition in and out of work mode. I have a forty-five-minute drive from Fort Collins, CO, to Cheyenne, WY, every day. The entire drive into work, I am ramping up and going over the day in my mind (and listening to my very random shuffle of music). By the time I enter the squadron building, and after trying to run up the stairs ahead of one of my assistants yelling "Fired Up!" before I get to the top of the stairs (it's a little competition/ritual he and I have), I am fully mentally prepared to attack the day. I say good morning to everyone in the hallway as I walk toward my office. Then I spend the first thirty minutes in my office with my door cracked (not as inviting as my usual wide-open door) as I review my emails and notes for the day. After thirty minutes I open the door fully and start walking around the offices for a more thorough beginning to the day with my people.

On the drive home, about fifteen minutes into the drive I hit the Wyoming/Colorado border. Since I am also a husband, dad, and football coach, I consciously force myself to stop work-related actions (even ruminating over work in my mind) and transition to the other part of my life I'm about to jump into at full speed for the rest of the evening. This also gives my people their evenings to do the same, without me constantly bugging them.

You might not have a forty-five-minute drive, or a physical state border in which to make this a little more measurable, however try to apply this principle to your work-life balance. It will make your life and the life of your people much more stable and avoid burnout.

Bottom line: everybody hates meetings, so don't make your meetings suck. Make them productive. Be brief, be funny. If you're not funny, at least be brief. Keep it under an hour. Don't try and cover more than three topics. Keep the meeting on topic by being the auctioneer to ensure you're getting input from the people smarter than you (i.e., everyone else at that meeting). Give credit where it's due, and most importantly, leave each meeting with actionables. Remember, what you all *do* once you leave that meeting matters more than what was said.

Chapter 12

Command Presence

Always be operating at least one rank above your current position, *no matter the rank.* Applying the principles outlined in this book can help you build the correct behaviors and mindset, *now.* Then, once you achieve the official position and title that goes along with being "the boss," all the things you've already been doing will be second nature. Plus, you will have the added authority to push past some of the bureaucracy you had to trudge through before you were promoted. I have been applying these same principles for the past sixteen years of service in the military, even when I was an E-3.

The next two chapters will feel like they lean heavily toward those already in a high-level leadership position. Hopefully up until this point, those reading that aren't yet officially "in charge" have been able to glean the intent of leading above your rank with the tactics, gouge, and war stories shared thus far. I want to stress yet again that no matter your rank, position, or career (within the military or civilian sector) these principles are applicable across the board.

The reason applicability is universal has to do with the natural human ability to influence each other simply by the way we live and interact. Influence does not have to be attached to a position and title. In fact, sometimes the most influential people in your organization reside in the mix of it all…mid-level employees with no supervisory duties. When you gain the respect of and buy-in from these employees, any change processes you attempt will go smoother, because these people have "street cred" and natural pull within the organization. If they say it's cool, then others will think it's cool as well. Really, not much has changed since middle school.

When I commissioned from an E-5 to an O-1E, I went from controlling some of the busiest airspaces around the world, supervising five Airmen, and being a Lead Instructor…to being the "Snack-O." The Snack-O is the Lieutenant (or a lower ranking enlisted member) in charge of the snacks in the breakroom. Not the most coveted or glorious additional duty within the unit. Nevertheless, instead of considering it a demotion in responsibilities, I revamped the hell outta the snack offerings, breakroom vibe, and slush fund accounting. It probably helped that I had four kids under the age of seven at the time, so my Costco snack game was dope (tip: become good friends with *Urban Dictionary* for your latest meaning of English slang being slung around by the younger employees).

Eventually, pilots from other squadrons would swing by just to get the latest snacks and breakroom vibe. I ended up earning so much profit off the snacks that we were able to buy custom retirement and going-away gifts that slap (again, please reference *Urban Dictionary*…or in-resident teenagers like I do). I wasn't too good to be the Snack-O, and I applied these same principles to make it excellent. Better believe this caused me to gain non-positional influence in my new work section as an O-1 Air Battle Manager trainee.

I also busted out the vacuum at least once a week to vacuum everyone's office (including the enlisted offices), as well as took out their garbage without being asked. It may sound silly, but it worked, because it reflected how I did everything else in my life. *How you do anything, is how you do everything.* So of course I also always wrote high-and-tight emails, had a squared away uniform, spoke confidently during any meetings, gave excellent tours to high-level officials, studied my ass off to be a master of my current craft, listened to fellow employees when they had to vent, provided gouge to others when they were struggling to find their own way, took care to not speak badly about fellow employees with anyone (except with my wife), and didn't make a fool of myself outside of work. Those things never changed and never will.

All this is to lead you into the "Command Presence" chapter. You can have a command presence without being the commander. You can command respect just by the way you walk into a room. So, as I review some of the ways I maintain a command presence (as a commander), try and be creative and contemplate how you can do the same in your own *current* position.

Command presence could be summed up by one of General Patton's famous quotes pulled from a 1944 letter he wrote to his son who was currently attending West Point: "You are always on parade."

Whether you like it or not, you are always being watched, observed, judged, and measured. Especially today with the ability to share information, you can bet your actions and words are constantly being recorded or talked about...which can then spread very quickly. I'm not talking headliners on national television or even on social media. It can be as simple as workplace gossip. You know what I'm talking about; it happens daily within every organization. No one and no organization are immune to it. So, let's give 'em something to talk about, how about lo... ...leadership (hopefully you don't get that play on song lyrics that just aged me a bit and gave you a peek into just how all-over-the-place my music shuffle really is).

Let them talk about how awesome the leadership is and what leadership did recently to impress them or make them enjoy coming to work. It could be something as simple as walking through the main building on base blasting "Danger Zone" on your speaker while wearing your "Iceman" costume on the annual "wear-your-Halloween-costume-to-work day." Okay, that's a little extreme, but you can guarantee afterwards people were talking about how the 243d leadership has fun at work. Just try and guide the inevitable workplace gossip by giving them positive things to talk about. A better example may be how you handled a tough situation during a heated meeting and were able to control the room without losing your cool. People talk about that stuff around the office, so make sure whatever they're talking about helps build the command presence you're trying to achieve. You do this by "always being on parade," or in other words, "always being on."

⋅◆——————————◆⋅

Although I am just a captain, I am the commander and the highest-ranking Airman in the squadron. It is protocol to call the room to attention when the commander enters for Commander's Call. I take military customs and courtesies seriously and believe in their intent to reinforce chain of command, discipline, and respect. However, I walk into my Commander's Call carrying a very large wireless speaker blasting walkup music, and then put the Hawks at ease after reaching the front of the room. It brings a fun mood to my meeting, while still upholding a very important military tradition of calling the room to attention for the commander. So there are ways to both maintain "positioning" while having a fun style.

"Positioning" can be done through many means. Obviously, walking into a room and having everyone stand for you is very overt and physically symbolic. Other examples of this in the military are positions in a conference room. The ranking officer usually sits at the head of the table. So, when it's my meeting, in my conference room, that's my seat and everyone knows it and doesn't sit there. Except one time my Logi (logistics and plans troop) whom I will refer to as "TSgt Worldwide" who likes to constantly mess with me (in professional ways) sat in my seat right before I came in to start a meeting. So I reached over and rested my hand on his shoulder just as Bane did in *Batman: The Dark*

Knight Rises, and in my best deep and slow Bane voice asked, "Do you feel in charge?" Once again, maintaining my positioning, but in a humorous way that reinforced a personable command presence. We have fun here.

[Note: After this example and within this chapter, is a good time to emphasize that even as the leader, you've got to be able to laugh at yourself and not take everything so seriously. Your demeanor will tell the room what level of seriousness needs to be maintained during a given situation, so definitely show your people all your sides so they know when you're joking and when you're serious. Don't always be so sarcastic that your people won't be able to tell the difference between sarcasm and serious…that just makes it super awkward for everyone involved.]

On the topic of "positioning," when I first took command and was attending meetings with fellow commanders, I was the only O-3 amongst a table full of O-5s. You better believe I took my seat at the table. I didn't timidly enter the room and sit along the edges. Why? Because I didn't do that when I attended enlisted meetings as an E-5 either. It's such a classic move, I'm sure, in most organizations, to see people timidly sitting along the edges or in the back of the room, until leadership enters and has to force people to come sit upfront. Sure, there are certainly meetings where seats at the table or front row are reserved for certain people, but if not, belly up! Yes, I am that nerd that sits at the front of class. So even in large auditoriums when most will sit in the middle or toward the back, I'm sitting toward the front in the first couple of rows. Let your presence be known.

A lot of this chapter has to do with developing your "swagger." I believe swagger is important. It's the non-verbals that gain you respect and set the tone. This doesn't mean it's okay to come off as a jerk. I've known many introverts that were effective at having their own unique swagger and setting the tone appropriately. As the leader, you set the environment for every meeting, every engagement, and every expectation. There are definitely times when I walk back to my squadron from a meeting on base, and just by how I walk down the hallway and set my face, my people know the tone I expect until I decide to change it again. It's okay for your people to see you pissed off or super elated. Go ahead and lead with your gut (different than emotions). Take control of your culture and drive your expectations with your current swagger.

The vast majority of the time, my swagger is going to come off as high energy, or most often referred to as 'Fired Up!' The number one compliment I get from people is, "I love your energy." They can tell how truly passionate I am about my job, and I don't even have to say much. Conversely, I have also been told I can come off as intimidating and abrasive. Most of the time when I am told this I am already fully aware. That was the point. I wanted to come off that way, because the situation warranted it.

However, I need to constantly be self-assessing to see if maybe I'm giving off signals that I wasn't intending...I do sometimes have RBF when I don't mean to. I've also been told that sometimes during meetings that I'm not running, when I start getting super bored or frustrated, my face and mannerisms show it. Apparently, I'm not good at hiding my "inside voice." If I'm not purposefully trying to give off this vibe (which sometimes I am), then I need to be aware of when I am doing that.

Whatever you want your style to be, just be purposeful with the messages you are sending to those around you. Remember, you are always on parade. So, look the part, sound the part, and act the part of leader. If you ever just need a break, close your door and consume some energy. Most days I am able to block off twenty to thirty minutes to shut my door, take my blouse off (the uniform jacket...so that I don't spill on it and look like a bag of ass the rest of the day), and eat my lunch in privacy. Although I'm often checking emails and texting while I eat, I am alone and out of sight. It's important to give yourself these types of breaks from being on parade. Reset yourself, suit back up, always have a ton of breath mints and gum available and get back at it!

As the commander, I want to be present for my people. I also want to be the one to represent the squadron during key engagements. Often the top leader in an organization is running from one thing to another and doesn't slow down for the more important "sand" of the day that slips right past them. This is why I have a policy regarding a few items I always want to be a part of. One of those that I take very seriously, is swearing in a new recruit or reenlisting one of my current Airmen. Technically, any commissioned officer can administer the oath. However, from the very first week of assuming command, I made it clear to my administrative assistant and the recruiters that I would be administering all oaths for those joining and currently in my squadron.

A phrase my DO at the time and I came up with was, "We pause for the recruiting cause." Luckily my DO, who had been there a couple years before me, had already developed a very solid relationship with the recruiters on base. So, I was able to slide right into those strong bonds and continue them after he moved on. Recruiters would usually schedule any visits from recruits and their families, so they became part of my planned sixty percent of the day. However, they also knew my open-door policy and they all had my direct cell number, so occasionally they were part of my unplanned forty percent of the day that just popped in.

Sometimes they would be walking a recruit around to other units and hadn't planned on bringing them by, but the recruiter asked to at the last minute. Following our mantra of pausing for the recruiting cause, I would drop what I was doing and engage with the recruit and their family. I love getting to know people, their story, and their "why." I would take as much time as they needed, sit and converse, show them around, get them a tour of the control tower, etc. When it came time to enlist that recruit (which I can't remember the last time a recruiter brought a recruit by to meet me, and that person didn't end up joining the 243d), I was the one raising my right hand along with them, reciting the oath and watching their families look on with immense pride.

Then after they complete BMT and their technical training, I welcome them back in front of the entire squadron during a Commander's Call, poke fun at them to let them know they are now officially part of the family and present them with our squadron patch. Within their first week on the job, I sit down with them, hand them my commander's packet and review my expectations (as outlined in Chapter 6). All these efforts, from the time they walked through my office doors with long hair and a beard, to crisply in uniform receiving their squadron patch, are showing them their commander is present. That I care. That I want to know their name and their story and be someone they can trust.

This also goes for any Hawk that is reenlisting, promoting, earning their next level badge, etc. As well as anytime someone earns an award or deserves a commander's coin, I reserve these moments to be present for and administer. I want my people to know I am proud of them, and I want them to see it. That's being present as a leader. Don't surrender those unique opportunities to someone else. Be there in person, for those moments in your people's careers (and more importantly, apply these same principles to your family).

You also need to be present during the hard work, when your people are "digging trenches" and doing the grunt work. I mean this figuratively and literally. My first experience setting up our equipment out in the field with my squadron was at a local dirt landing zone. Our medium and heavy ATC packages require a great deal of manual labor to set up. As soon as we arrived at the site for set up and had our pre-mission brief establishing expectations, it was time to execute. Since it was my first time, I spent the first fifteen minutes observing the action and my people getting after it. Once I had seen a solid overview of how the operation was being carried out, I picked up a sledgehammer and started hammering radio antenna support wire stakes into the hard Wyoming dirt.

Next, I helped raise the mobile navigational aid with the hand crank. Then I moved onto pounding grounding rods three feet into the dirt to establish the correct ohms needed to run the support equipment generators. Lastly, I helped set up where we were going to eat our sack lunches by shoveling cow manure out of the pasture where the tables were being placed. After lunch (and cleaning the shovel), I challenged a few of the Hawks that were throwing a football to a contest of who could hit the football the farthest with the broad end of the shovel. It was a blast. That day we sweat together, ate together, and played together. It doesn't get any better than that.

You are never too good to shovel shit.

You want your presence to always mean something. As much as you need to be personable and likable, you've got to be careful not to become just "one of the guys." You need to maintain the expectation that what you say, write, and do…needs to be heard, read, and understood. If you talk too much, send too many emails, and are always "just hanging out" in the shop…then you become "one of the guys," which is different than being "a relatable leader."

When someone sees your email in their inbox, it should carry weight and mean something. You don't want them to "get around" to checking on what you sent them. Be especially selective with how often you send "Squadron/All" emails, and when you do, keep it short. Go for impact. Again, this applies to anyone wanting to be a better leader, even if you're not in a leadership position. Don't become irrelevant to people because you send a million emails out a week or tell really long stories. Emails and long stories are like poetry, and most people hate poetry. Don't become white noise.

•◆————————————◆•

Lastly, be decisive. Developing a strong command presence includes making unwavering decisions. This does not mean you make rash decisions meant for speed and not accuracy. All thrust and no vector only puts you in the wrong place faster. That is why the way you conduct your meetings, as mentioned in the last chapter, is crucial to the success of your decision-making methods.

Surrounding yourself with smart advisors who have the best information will allow you to develop COAs from which to choose. Once you have carefully balanced the intricacies of each COA, be decisive with how you choose which direction to take. This does not mean you keep driving down a road that you eventually discover has a cliff at the end. You need to make adjustments when absolutely necessary. However, they should be small adjustments if any, because you frontloaded the COAs with smart people and sound advisement. So be confident with your decision, even if it's only a seventy percent solution (or fifty percent plus one, if you're General Patton...more on that in Chapter 14). Your people will mirror your confidence...or lack thereof.

One example of this was when I secured the F-16s from South Dakota for Triple ACE. They were the aircraft we had chosen to conduct the HPR (Hot Pit Refuel). I had decided, after being advised of all options (and coached by a Weapons School graduate), to use the HPR event as my "timeline hinge." In other words, every event taking place before and after the HPR was built off that target HPR time (and we had over eighty time-targeted events every day of the exercise).

To further illustrate, this "timeline hinge" even determined when the F-16 maintainers in South Dakota would show up to work, so that the F-16s flying to Cheyenne that day would land at the exact planned time (Time on Target – TOT). Which also meant it drove what time the C-130 would land that would then get fuel pulled off its wings (Wet Wing Defuel), and when the fuel truck would be ready to offload said fuel, at a certain detailed pump rate. A domino effect as you can see.

When the runway construction company updated us a few weeks out from the beginning of Triple ACE that construction would not be completed in time, we had to sideline the F-16s and replace them with the UH-60s. Since we had planned for this contingency, we had already sent the fuels bubbas to train on the UH-60 and built a supplemental timeline adjustment. This adjustment did not phase my confidence, because we had planned for it.

Something I *hadn't* planned on, as already mentioned, were the CH-47s getting grounded nationwide the night before Triple ACE kicked off. Less time to problem solve, no contingency already built-in, and much more at stake. Did I quiver and waiver when I received the news? Did I pull my staff into my dorm room and show defeat and worry on my face? Negative. I relayed the info, paused for everyone to take it in, and then confidently commanded my team. They rose to the occasion and with only my intent, they developed the plan that saved Triple ACE and ended up working without a single hitch. I was so proud of them, and all it took was them seeing from their leader that everything was going to be okay, and that I had confidence in their abilities.

If taking care of your people is the foundation for your "why" behind leading, then your command presence will flow naturally. Find what works for you, develop your own genuine swagger, and do what feels natural. Lead from your heart and gut.

Chapter 13

Commander's Intent

How many times have you left a meeting and been confused or not known what you're supposed to do now? Then you slide over to a trusted colleague and whisper, "So...what exactly was that meeting all about and what does the boss need us to do?" I know I keep emphasizing how this is the leader's fault, which it is. However, I will also portion some of the responsibility onto the employee since they should've asked the "stupid questions" before leaving the meeting so that they weren't still confused.

This is why I took an entire commander's call just to teach my people about "Commander's Intent" when I first took over, and why I include it in my Commander's Welcome Packet for new Hawks.

While I was an OTS Instructor (I know, "Please not another OTS story!") one of my additional duties was Lead Fitness Advisor. Basically, this meant I was in charge of the first and last PT tests administered at OTS for graduation. I also worked with the enlisted PT Leaders (PTLs) on executing these tests, as well as developing the group fitness workouts throughout the nine-week course. This was no small task, since most classes were around three hundred officer candidates, and I was expected to get them all tested on push-ups, sit-ups, waist measurement, height, weight, and 1.5-mile run...in less than two hours.

I was then required to have all their scores uploaded for the commander to review that same day.

I took over the position from a departing Instructor and received the gouge on how things were being done. It took no less than my first time observing how this was normally done, before I knew an overhaul was going to be needed. Especially considering that two classes away we were running both training squadrons congruently (we usually staggered them a few weeks apart) and each squadron would have almost four hundred officer candidates. This meant nearly eight hundred trainees needed to be tested in the same amount of allotted time on the schedule. This specific summer of 2019 Air Force OTS class would come to be known as "Godzilla."

On an early, foggy, rainy, humid, and hot Alabama morning, I observed the first PT test process. I was not in charge of the program yet, just observing in preparation to take it over. The squadron commander was also present. As the rain kept coming down and trainees started conducting their push-ups on the wet steamy pavement, the process began to deteriorate. Train wrecks started to occur between stations, trainees were standing around waiting when they should've been conducting the next exercise, there weren't enough sit-ups pads, and Instructors were starting to lose their military bearing in the heat. It eventually got so bad that the commander pulled the Instructor in charge off to the side and chewed their ass.

Cool. I was really looking forward to taking over this hot garbage of a program.

The next day the commander called me into his office, wanting to discuss how we were going to improve this process and get it down by next class, so that by Godzilla it would be perfected. I showed up with my notepad and pen ready to problem solve.

During the first ten minutes the commander rattled off a lot of stats. He was a data-driven leader (not my favorite) and was crunching all sorts of numbers: transit times between stations, total number of trainees multiplied by number of seconds it takes to test each element of the PT test, the average speed at which an Instructor could type the results into the database for official scoring purposes, etc. My non-data-driven brain was spinning, as I was frantically writing as many notes down as I could.

After ten minutes of this I found a tactical pause in his talking and interjected, "Sir, what is your intent?"

He paused, looked at me, and said plainly, "I want eight hundred trainees tested in less than two hours." Boom. Got it.

Next question, "Sir, how many Instructors will you permit me to pull from standby in order to get this accomplished?" (standby or "support" Instructors were those not actively teaching or "pulling" a flight of trainees; we took turns rotating each class). The commander replied that I could have as many Instructors as I needed. I then asked if he had a preference regarding the location of the test, because I had some ideas on how to change it up to be more efficient. He said he'd entertain ideas. From there I exited his office and got to work.

Two days later I had two different COAs to present him. He chose one and tweaked it slightly. I took the adjustments and last requests of the commander and ironed out my final proposal. It was a total revamp of where and how we conducted the testing. It would require moving the start and finish line of the 1.5-mile run, conducting the height/weight/waist measurements inside the dojo warehouse, and performing the push-ups and sit-ups on the dojo's large covered artificial turf area. Essentially, all six components of the test would now start and end in the same location, whereas before they were scattered between the OTS gym and the outdoor track. Additionally, it eliminated the amount of equipment needed since sit-up pads were replaced by the soft artificial turf. I also created a scoring table which fit all sixteen members of a flight along with all of the necessary scores, onto a single page.

We gave it a trial run during the next regularly sized class, and it was magical both times (beginning and end of OTS PT tests). Confident in my abilities, the commander blessed the process for the Godzilla class.

The first PT test at the beginning of Godzilla we successfully tested 783 trainees in 1 hour and 57 minutes. By now the Instructors had followed my new process three times, and we aimed to break the time record during the last PT test of Godzilla, which we did, in 1 hour and 48 minutes (oh, and I wasn't offered $1M to find a way to make it happen). That summer we ended up graduating the largest Air Force OTS class since the Vietnam War, and they were fit-to-fight.

This story helps illustrate exactly what commander's intent looks like (or "intent" in general, regardless of who is giving it). At first the commander was getting too far into the weeds. He was not taking the advice I later received, to only do things only a commander can do. He was essentially doing my job for me, by crunching numbers and developing COAs, and he doesn't have time for that. So, I helped guide him directly to his intent, which was "Test eight hundred trainees in less than two hours." From there I prodded for a few resources and permissions, and then exited his office and started to take action on his behalf. Since I took notes, I could easily repeat what it was he said and wanted, as well as relay to the Instructors and PTLs what it meant.

It would have been more efficient if the commander had just given me these things from the beginning, but it was a good lesson for me to realize that sometimes when you're too close to the issue and focused in on the details, you forget to zoom out and keep it simple by framing the desired end state.

From this experience I was able to develop what commander's intent was, how to check myself to ensure I'm keeping it simple to understand, and lastly how to teach it to all of my Airmen. This is Commander's Intent simplified, and then explained:

Commander's Intent

1. Repeat it

2. What does it mean?

3. ACTION on behalf of the Commander

Repeat it. In order to even attempt to explain something that someone said, you need to be able to repeat it. Maybe not verbatim, but not off the top of your head either. Success or failure rests on this first step! How many times has something gone wrong simply because someone repeated the communication slightly incorrectly? We all know the "telephone game" and how that analogy applies here. Therefore, to keep this entire process from breaking down from the very first step, BRING A PEN AND PAPER TO EVERY MEETING! Write notes down to help you remember exactly how it was said, or at least jot down the key phrases or major highlights. This gives your brain moments from the meeting to point back to, instead of digging into the depths of your nugget to just slightly screw everything up. Don't be a hero and try to remember everything or think that you "got it." Even if you do have some amazing gift of memory (which you don't), you will also look more professional and attentive to the person talking if you occasionally glance down at your paper and scratch a note down really quick. Can you see how taking notes on your phone could send the wrong message, even if that's what you were actually doing during a meeting? So bring a pen and paper and take good notes so that you can repeat whatever main objectives were talked about.

What does it mean? Good job, you showed up and took good notes. Now, with minimal glances at your notes, could you walk out of that meeting and explain to a random person walking down the hallway what the commander just said? This is where asking questions during the meeting is *super* crucial. I am the questions guy, I don't care about looking dumb, because guess what...I don't miss deadlines and my products are high-and-tight. Always. You can go ahead and look super cool not raising your hand, but when you give the commander a subpar product or miss a deadline, now who looks dumb? So, ask the questions then and there. Not only does this help you and others in the meeting better understand what is meant, but the commander (or leader of whatever meeting) appreciates the engagement. It helps them think of other things they may have forgotten to mention, or it at least gives them insight into what's going through the minds of those in the meeting. I don't want to just dictate and hear myself talk. I want input and questions, and I want people to poke holes in my plan so that it can be the best version. Therefore, after writing it down and being able to repeat it, ask questions until you're comfortable enough to tell someone who wasn't in the meeting what was meant by what the commander said.

Action on behalf of the commander. Now it's time to apply the "What you *do* out there, matters more than what you *say* in here" quote. Take action! But not just any action—action *on behalf* of the commander. You are now representing the commander after being given their intent and permission to execute. Now you must *do* something about the notes and understanding you took away from that meeting. It means nothing otherwise. It is also empowering to know that you have the backing of your commander to do whatever it was they said they intended to have done.

Moreover, commanders and leaders or supervisors, after you give your intent to a subordinate, then it's time to provide the "top cover" for when mortars start falling from the skies as they carry out your intent. Protect them, empower them, back their play. After all, they are doing the things you asked them to do (unless they didn't follow these steps properly and are way off course, then it's your job to vector them back on course...remember, "it's your fault!"). Otherwise, if you don't empower and protect your people, and allow them grace after failed attempts, they will stop doing things excellently for you. Why would anyone work hard for someone that turns and runs when something goes wrong? Don't be a spineless leader.

◆————————◆

Transparency is a critical ingredient for clear communication. I am willing to share plenty of information, especially if you are asking genuine questions intended for clarification. This doesn't mean I don't know how to protect classified information or that I divulge things given to me in confidence. Rather, I like to give the command-level perspective and information to anyone needing it to perform their jobs.

Why are humans unwilling to share? Does it make you feel more powerful to have information that others don't, and then clutch onto it to make it appear you are more special than you really are? Stop hoarding all the information. If by being more transparent you're afraid you'd become irrelevant as a leader, then you already are. Better access to information doesn't make you a leader, your actions do. And since your people are the ones acting on your behalf, they should probably have the majority of the information necessary to get it done properly.

In over three years as a squadron commander, I have yet to get in trouble for sharing too much information with my people. Instead, I have witnessed employees excel well beyond their previous performance levels and take ownership of their work. I wonder if the two are correlated?

It is your job as the leader to be very clear about how you communicate your intent. However, not to contradict what I just wrote by making this sound like you need to micromanage or spell out step-by-step directions on how you intend something to get done. To combat inadvertent micromanaging, focus on *what* you want to get done, and empower your people with the *how*.

Remember that I wrote back in Chapter 5 about our squadron TDY to Fort Carson to all qualify on our weapons? What I didn't tell you was that our departure date from Cheyenne for this TDY fell one day after the largest two-day snowfall on record for Wyoming: thirty inches. The location we were intending to drive (Fort Carson) was a few hours' drive down to Colorado Springs, CO, just south of Denver, which had received twenty-seven inches of snow. This blizzard became known in that region as "Snowpocalypse 2021."

It was the school district's "spring break" for my wife and four kids. When we saw the forecast coming, the five of them took the four-wheel drive SUV and hightailed it away from the storm to stay that week with extended family...leaving me with my small front-wheel drive commuter. Knowing the storm was coming, I pushed the start of our TDY one day to the right, to give those flying in from out-of-state time to catch delayed flights, as well as those of us in town to dig ourselves out.

The entire Air Force base shut down during that week after the storm, however we needed to get on base to dig out the buses and get our equipment out of our warehouse, not to mention have a place to park everyone's vehicles. However, no other squadron was going in to work for at least a week. Power was out all over the place, parked cars were simply invisible—covered in snow and inaccessible, and main roads were not even close to being all the way cleared. It was an incredible amount of snow, and nobody wanted to dig out of it.

Commanding from my living room, I orchestrated my intent. I made it very clear to my leadership that we were *not* canceling this TDY, and a one-day delay was all I was willing to budge. It became a test of our ability to communicate with and maintain accountability of our people from a distance, our resolve to find a way no matter what, and a golden real-world experiment for me to gauge my own willingness to assume higher levels of risk. Snowpocalypse provided us a huge opportunity to gain strength and experience.

The days during the blizzard were intense. So much coordination had to occur to ensure we maintained proper accountability of our people and their safety. We had sixty-four Hawks proposed to attend this specific TDY and nearly half of them were coming in from out of town. The next twenty-four hours consisted of phone calls and texts back and forth, updates being kept on notebook paper and pictures of said notes sent via text to those without computer access. Reports of delayed flights from all over the nation from the DSGs flying in, pictures pouring in of snowed-in doors and cars, calls about Hawks taking alternate routes on highways to zig-zag their way to Cheyenne, and neighbors helping each other dig out their cars. One particular Hawk skillfully negotiated with half a dozen state trooper checkpoints that weren't letting anyone through…and yet he found a way.

As all of this was going on, my intent remained clear: everyone will arrive on time to the base in Cheyenne, on this date, at this time...period. Find a way, there are no excuses (another, "If I gave you $1M" moment). This wasn't because the mission (qualify on the M-4 and M-9) was do-or-die. It would've been very inconvenient to reschedule and find a way to get us all our annual qualification, but it wasn't going to cause a huge readiness issue, plus I had the best excuse in the world to cancel and not get any heat. Nevertheless, not getting any heat and playing it safe wasn't the point.

I wasn't going to let a blizzard prevent us from exercising our capability to be the Minutemen we claimed to be. If the National Guard couldn't dig themselves out of a blizzard, then who else was expected to come to the rescue when the call came? If not us, then who? So, it became a momentous chance to practice being the dependable less than one percent, during a real-world crisis. I wasn't going to let this opportunity slip by, and that is why I increased my acceptable level of risk.

I, too, had to take breaks between phone calls and texting in order to dig myself out. To make matters worse, only the main roads inside my subdivision had been plowed, and we live on a cul-de-sac. So they ended up inadvertently plowing an even taller wall of snow perpendicular to the exit of our cul-de-sac that was taller than my car. I literally had to shovel a path just wide enough for my small commuter car to back out of my driveway and down the cul-de-sac road to the main subdivision road. It took *forever!* Once I had finally shoveled a path to the wall that was taller than my car, I ran out to the main road and waved down one of the skid steers plowing away. I pulled out a twenty-dollar bill and asked him if he'd knock down my wall. He took the money and gave it a couple solid rams to break through, and then I finished clearing a path.

I also live on a hill, so once I felt I had a path, I attempted to make it down my cul-de-sac road and then up my main subdivision road. My first two attempts I ended up getting stuck and had to wave down trucks to pull me out. The next day was when I had established all of us to meet up on base, so I had to make this happen. More hours of shoveling, between coordinating with the other Hawks doing the same thing and flying/driving in from out of town. Finally, I was able to break through and make it up my hill to the freeway. I drove up and down the freeway to ensure I could make it the next morning. Good to go.

My DO wasn't so fortunate with his subdivision getting plowed in time. So instead, he had to build makeshift snowshoes out of whatever he had in his garage, to then trudge through a half mile of waist-high snow while carrying his luggage, just to make it to a plowed main road where some of our Hawks were able to then pick him up.

My Chief lived off the grid well out of town and had to use his tractor to plow a path to the main road, all without power and having to conserve his cell phone charge.

A couple Hawks carpooling from out of state had to sleep in their cars halfway to Wyoming because all the hotels were full.

Another Hawk, instead of waiting for her delayed flight to get rebooked, rented a car and drove to a different airport that had a flight that would get her there on time.

Story after story of Hawks getting it done kept flooding in. All the while my staff and I kept track of where everyone was and their safety status. It appeared this was going to work...then I remembered the base was closed due to the storm! So, I called the Security Forces Commander to ensure the gate would be unlocked, the Civil Engineering Commander to ensure the roads and parking lots to my squadron would be plowed, and the Logistics Readiness Commander to ensure an Airman would be present to open the warehouse and issue us our gear. It was super impressive to behold these other three squadron commanders come together and orchestrate their pieces of the puzzle for our success.

After it was all said and done, sixty-four of sixty-four Hawks showed up on time for this annual training event.

You know the rest of the story. We had an absolute blast down at Fort Carson, qualified on our weapons, and made memories none of these Airmen will ever forget (to include renting an entire indoor paintball arena for the whole squadron the last day to exercise our newly honed marksmen skills on each other…with my DO and I allowing each Hawk to take a free sniper shot at us at the end). The Red-tailed Hawks made it happen. I gave them my very clear and unwavering intent, and they found *how* to make it happen. It also showed me the mettle of my people and gave me the confidence to challenge them with more in the future; and every time I have, they've risen to the occasion. Fired Up!!!

That may be an extreme and oversimplified example of commander's intent, but it proves how simple this process can be and still not implode. Relating this leadership tactic to more day-to-day operations, you can always be very clear with how you communicate anything by putting the words, "My intent is…" at the beginning of your sentence. For me, when I phrase it that way, it helps my brain cut the crap and get to the point quicker. Forcing yourself to put it into a sentence that starts with, "My intent is…" (like I had to do for my OTS commander during the PT test revamp meeting) may just help you filter through the data and pull up from the weeds.

◆———————————◆

Sometimes you will need to put your intent into writing and sign it. We call these Policy Letters in the military. This is one of those authorities you have as a G-series commander to create, so long as it's either amplifying a current regulation, or making it more restrictive. You just can't make a regulation less restrictive, i.e., allow facial tattoos when the reg says no facial tattoos. I have leveraged this authority a handful of times. I have written policy letters concerning squadron specific uniform wear, medical reporting procedures, delegation of signature authority while I was TDY, squadron awards program rules, and policies specific to our quarterly drill privileges.

The one that probably best demonstrates outlining commander's intent through a policy letter rather than just outlining existing rules, has to do with manning our full-time ATC Tower and RAPCON (Radar Approach Control) over drill weekends. Due to our full-time control of the Cheyenne Regional Airport being manned by guardsmen as well as civilian DoD controllers under my command, those uniform wearing controllers need to be with the rest of us during drills. However, due to a lack of civilian DoD controllers and crew rest restraints, there were times when we had to utilize military controllers during drill. So I outlined in a policy letter the "why" behind needing military controllers attending drill, the "why" behind different statuses of employees (full-time vs. part-time, and military vs. DoD), and then ended with the progressive courses of action that would be taken by the facility managers in order to fill the drill weekend schedule.

It specified my overall intent, and then outlined three steps that should be taken in the order I wrote them, to best meet that intent. That way, everyone can refer back to it when they have questions. It saved me from having to repeat myself, and it also ensured my intent was being met as best as possible. Although I did provide some of the "how" through that policy letter, I still empowered the facility managers to overrule the progressive order if they had solid justification. At the end of the day, my intent was to get as many military controllers to drill as possible.

If you're still struggling to find which words go after "My intent is…" just take a step back and try to remember what it was like not being the leader. Remember back to those glorious days when you could complain about everything, because you didn't have any responsibilities? Oh, those were the days. You knew so much, and leadership knew so little. You had all the answers while standing around the water cooler with your other ignorant friends, complaining about this and that, and scoffing at how senseless your supervisor was. The best were those meetings when you'd snicker in the back of the room with one another as you whispered things back-and-forth that you now realize were naive oversimplifications…

The serious point is, don't lose your perspective as a leader. Don't suddenly fall into the traps you witnessed so many poor leaders before you blindly trip into. Remember how simple it really is to motivate people and take care of them, which in turn takes care of the mission.

Don't add unnecessary fluff to your intent or meetings, and occasionally try listening to those standing around the water cooler. Sometimes they have some solid ideas. Other times they are just blowing off steam. In rare cases, they are just complete morons and make up your bottom ten percent that every organization has, and you need not waste your time with them. Just remember what it was like to "be a kid" and try to keep your intent simple enough for everyone to follow and empower those with the good ideas to find *how* to make your intent a reality. You know what people want because it's what you always wanted out of a leader. Now it's your turn to be that leader.

◆———————————◆

"Generate F-15 sorties." This was the mission statement the new Mountain Home AFB Wing Commander created when he took over while I was stationed there. I absolutely loved it. It was, and still is, the only Wing-level mission statement I remember. It was so simple. No matter what your job was on base, you understood that the base existed to train F-15 pilots for war, and you had a part to play in that mission. Your part may have been very indirect, but with some creativity you could follow the path from your job to a pilot taking off. It brought everyone into the fold of purpose.

Even some of the gate guards started saying it when I'd ask how their day was going as they saluted coming through the gate, "Good, sir. Just generating F-15 sorties!" They may have been saying it to mock, since they might have felt very far down the line from influencing an airplane taking off. Or maybe they truly did see the importance of their job to guard the gate, in relation to training F-15 pilots. Regardless, the Wing Commander had successfully gotten his message all the way to the lowest ranking Airman checking IDs at the gate. Moreover, that Airman could now take ownership of their piece of the puzzle if they wanted to.

When I took command of the 243d ATCS, I asked where the mission and vision statements were located. It took some time to scrounge them up and dust them off; the first sign that nobody had a clue what they were. Secondly, when I read them, even I was confused…and I had been in ATC for many years. I won't repeat what I already wrote back in Chapter 6 concerning the new mission and vision statements, because the point I'm making in this chapter is that you need to leverage your authority to write your own mission and vision statements and use them as your overarching commander's intent.

I really liked what the principal of my daughters' elementary school said concerning their mission and vision statements at the ribbon cutting ceremony of the new school. Since it was a new school, she had the privilege of writing the mission and vision. She said that while she was developing them, she switched out the word mission for *purpose*, and the word vision for *promise*. It helped her better frame the words she chose. I thought that was a great way to distinguish the difference between the two.

However, while writing this book, I tried to find their mission and vision statements on their school website and couldn't. This leads me to my point about constant reminders regarding your baseline intent. I framed a total of sixteen professionally printed large posters of the mission and vision statement I had created and plastered them all over the walls of each of the four buildings our squadron owns on base. I also required it to be one of the questions on every promotion board. I know it's working because I hear my people weave it into their conversations when they're talking to other people outside our squadron. Make your intent known and keep repeating it!

Another tactic I use to ensure my people are constantly aware of my intent, is to issue what I call a "Commander's Charge" every month during my Commander's Call with the entire squadron. These usually are only a few words of action displayed on a single presentation slide with a related picture or graphic subdued in the background. Most often they have a story attached and are a way for me to tailor operations to the current squadron focus. I love putting these together and presenting them. I try my best to keep them brief. I also include the single slide in our monthly newsletter that is published a few days before drill. This gives everyone a chance to start thinking about what I wrote before I have a chance to present it.

I would recommend that if you don't have an official monthly gathering with all your people in one room, you should attempt to make that a regular occurrence. Even during COVID, we held a few virtual drills, and you better believe we were all on our computers together each day getting stronger, and that included me addressing a very large amount of small screens on my computer at the beginning and end of each drill day. Find a way to make this happen. Your people need to hear from you but make it brief and funny. Please, at least make it brief…otherwise, maybe don't do this. Also, if you're not THE boss, then hold these types of gatherings for the group of people you *are* in charge of.

After spending a year in command and having gotten a solid feel for the position, I started looking for ways to push my squadron even further. I saw so much potential in the people and the future mission. I started to develop a plan for progressing us faster and in a more focused direction. It was after that Weapons System Council in March of 2021 when I decided it was up to us to create new capabilities. I spent the next few months researching the extent of my authority, reading up on what top Generals were inferring would be our next areas of operation for war, and asked my people a lot of questions about where they thought we could go if there were no limits. I ended up developing a "State of the Squadron" address that I presented at our large August quarterly drill (I have since carried on this tradition and give a "State of the Squadron" address every August; a reflection on what we've done the past twelve months and where we're headed.)

This specific drill was a wakeup call for my squadron. Although we had been performing well and had started moving in the right direction, I felt like we were in third gear when we needed to be in fifth. So, I reviewed a lot of things with my Hawks over the course of a couple days by splitting up groups and lectures. Topics ranged from equipment and exercises to authorities and expectations. I made it very clear that their commitment and my willingness to keep them around were both voluntary, not obligatory. I showed the "What is your profession?!" scene from the movie *300*, gave personal examples of how my "why" had evolved over the years, and also showed clips of the 9/11 terrorist attacks since the next month would commemorate the twenty-year anniversary. It ended with a culminating Commander's Call where I reminded them of the Commander's Charge to "Create Capability" that I issued to them in April of that year (referred to in Chapter 7).

I summarized it all by explaining that my number one need as a squadron commander was mission ready Minutemen. I outlined it on the slide as, "CC's #1 Need: Mission Ready Minutemen." That's it. If you are ready by the Air Force's standards, and also personally ready (mentally, physically, financially, and familial), then you have been hard at work, because it takes a lot of effort to get an Airman one hundred percent deployment ready. This need, in conjunction with the mission and vision statements I had created, were the baseline commander's intent I still repeat to this day. Chapter 7 already outlined all the successes that came from my people taking these charges to heart and excelling well beyond my expectations, but it started with commander's intent.

As of the writing of this book, General Charles Q. Brown Jr. was the Chief of Staff of the Air Force (CSAF). *[Note: As of the publication of this book, General Brown had been promoted to Chairman, Joint Chiefs of Staff (CJCS)].* He is the General who authored "Accelerate Change or Lose" in August of 2020 (just months after I took command) and his charge gave me the confidence to act boldly on his behalf. More recently, he also published the "Air Force Future Operating Concept

Executive Summary" in March of 2023. I want to share a paragraph from this document that I believe embodies exactly what commander's intent truly implies:

"Mission command is a philosophy of leadership that empowers Airmen to operate in uncertain, complex, and rapidly changing environments through trust, shared awareness, and understanding of the commander's intent. When empowered in this way, Airmen do not have to wait for orders from the higher headquarters to make bold decisions and take advantage of fleeting opportunities. This type of leadership does not just happen. It takes intentional development and practice."

This type of leadership definitely doesn't just happen. Words he used such as *trust, shared awareness [transparency], empowered, bold decisions, intentional development* should not just be skimmed over. This is some heavy hitting leadership from the Air Force's top General (now the entire armed force's top General), and he is telling everyone below him to be bold! He is giving us all permission to do what is necessary to be the strongest military in the world! This is the type of commander's charge that gets me 'Fired Up!' and is what I try to emulate with my squadron of Red-tailed Hawks.

I tell the Hawks the *what*, and they figure out the *how*. I ensure I know what my desired end state is before I give my intent, but once I've established the intent, I let the experts do what they do best. Creating intent is a very important and serious responsibility of the person in charge. Don't take it lightly, and don't let your people fail because you were not clear and simple enough. Also, provide them the safety net to stumble into and don't condemn them for tripping up. Instead, *commend* them for their courageous efforts, dust them off, and send them back into the fight. Learn from your own mistakes as the leader and make adjustments when necessary. Most importantly, lead with intention and not by default.

Chapter 14

Relentless, Rebellious, & Rogue…
v Reckless

At the time of writing this book, the United States of America is only 247 years old. Think back to your history classes in high school and remember how quickly you breezed through historical timelines—247 years is about a quarter of the time that the Roman Empire held their dominant reign, and we cover that period in history within about a week in school. We must be careful as leaders within the U.S. not to let our powerful rise eventually fall into the "one hit wonder" chapter of the world's history book. In order to prevent this from happening, leaders need to be relentless, rebellious, and rogue. The United States of America rose to become a powerhouse country in part based upon these three character traits being embodied by our forefathers and many other historical American leaders we've had since (read about them and you'll find they share these traits). Unfortunately, these types of leaders are now uncommon.

To regain the dominating ways that paved our path to becoming the leading world power we…are? (…once were?), we need leaders in business, politics, law, defense, and the private sector to regain their damn swagger. The social climate we currently find ourselves in threatens anyone who leads from their heart and gut. It prescribes a narrow road to drive our words and actions down, careful not to hit any "career ending curbs." This is absolute bullshit and is deteriorating our American culture. That's not what made us a strong nation to begin with, and it's not what's going to keep us on top. When other countries see us pandering to political correctness, they see weakness. On my Air Advisor deployment alone, I interacted with people from a total of forty-one different countries for over six months. U.S. weakness due to political correctness was a consensus sentiment shared by the vast majority of those countries' representatives.

America needs to be the relentless, rebellious, and rogue sons-of-bitches that all other countries inescapably respect; economically, militarily, and politically. I'm not even saying we need to "go back" to anything, or that we shouldn't evolve. Our past is riddled with horrible mistakes and black eyes; that is why we shouldn't proceed recklessly. We need to learn from the past, gain strength from the new ideas of the present, and march forward into the future with resolve to not self-destruct by picking each other apart. We need to unite under one simple objective: maintain independence from government control of our lives (both foreign governments, and our own). Isn't that what started this whole "American freedom" thing anyways? Remember that document called The Declaration of Independence?

There are many strong passages in The Declaration of Independence, some of which are points of debate between political parties to this day. My objective in leveraging the mentioning of this document in this chapter is not meant to further the divide we currently need to mend. The purpose is to draw our attention as American leaders to what we all have in common: a desire to remain a strong and stable nation. One passage reads, "Prudence, indeed, will dictate that governments long established should not be changed for light and transient causes..."

I view many of the current issues we find ourselves "wound around the axle of sensitivity" with, as "light and transient causes." However, I truly believe we are capable of resolving these issues amongst ourselves. I by no means side with any other course of action regarding these matters, than that of internal democratic and peaceful resolutions. In other words, we don't need another revolution or civil war, y'all, so calm yourselves down and remember, we're all Americans and are on the same side at the end of the day. Let's focus on battling *the enemy without* and put our petty differences aside from within. As leaders, it's going to take us leading from our heart and gut to bring us back to the place we need to be as a country. Find it in your heart to love your fellow Americans and listen to your gut to focus on the common enemy.

◆———————————◆

I want my passion to come through during this chapter regarding these three strong words: relentless, rebellious, and rogue. Can you guess what all three of these words have in common?

Risk.

Without risk, you are simply routine, and routine is for robots.

My definition of risk: bending the rules or going against the odds *after* knowing the associated potential consequences. Before you go thinking I'm the guy that secretly breaks the rules and tries not to get caught, I don't mean you embezzle money from your company and risk going to jail. This is called *illegal*. I am not advocating that anyone do anything illegal. I am, however, promoting "operating in the gray" as opposed to playing it so safe in the black-and-white that you never push yourself or your organization to achieve greater potential. I won't deny operating in the gray, it's no secret.

A very rudimentary example of operating in the gray is not changing the engine oil in your car at the prescribed mileage. You run the risk of hurting your engine overtime if you decide to push past that mileage mark. Bending this by going a few hundred miles over likely won't do any damage, or maybe it will. This is a gray area, and we operate in these daily.

However, when it comes to leadership, most leaders are going to be risk-averse. Why? Because they are afraid (see Chapter 4: "If You're Scared, Step Aside").

In order to help you be more confident being relentless, rebellious, and rogue, you need to have your "why" always selflessly grounded in your people (see Chapter 5: "People First, People Always"). That is why I led this book with the chapters I did, so that you could understand the follow-on principles I would later write about. Having the underlying reasoning behind operating in the gray, grounded in the principle of taking care of your people, will prevent the majority of your decisions from turning into regrettable reckless ones. However, once you cross the line of "going gray" due to selfish or illegal reasons, you have lost what it means to be a leader and should be removed from your position.

If you can genuinely state and prove that your decision is founded upon taking care of your people (which in turn takes care of the mission), then you are ready to take risks. What chance of success percentage gives you the confidence to take the associated risk? It will be different for everyone, for every situation, and will evolve over time. However, referencing back to General Patton's letter to his son at West Point, I wanted to share with you what one of our most effective WWII Generals considered an acceptable percentage:

"Decide what will hurt the enemy most within the limits of your capabilities to harm him and then do it. TAKE CALCULATED RISKS. That is quite different from being rash. My personal belief is that if you have a 50% chance take it, because the superior fighting qualities of American soldiers led by me will surely give you the extra 1% necessary."

So according to General Patton the percentage for chance of success to then assume acceptable risk is fifty-one percent, because he believed his soldiers would tip the scale if all else were even. Where have we gone since WWII as leaders? Not just in the military. Whether you're a leader on Wall Street, at Walmart, or leading a charge on the Great Wall of China, are you willing to proceed at a fifty-one percent chance of success? Depending on the situation, that may fall into the reckless category, so I'm not saying this is my percentage for every risk assessment. The percentage value really isn't the point here. The point is *our attitude toward risk* as leaders. I believe we are more concerned in today's climate with who will be offended or how a decision will impact our personal careers, rather than being concerned with the necessity of said action to take care of our people and organization. We have changed what right looks like, based largely on perceptions, not reality. We have falsely changed the definition of success to be keeping your job by playing it safe and pandering to social pressures.

It's time we focus on taking more risks so we can strengthen our position as a nation in the world.

Our military enemies, business competitors, and naysayers want us to be risk-averse. They win if we hesitate on a solid plan because it isn't one hundred percent fail proof. They win if we delay due to worry. They win if we pander to the hurt feelings of every person with a social media account. This needs to stop. Everyone's feelings will be hurt when they no longer have those freedoms of expression because ours or another government has strangled these liberties. I appreciate that we have the opportunity to express ourselves this way (that is after all, what I dedicated this book to). However, that's not what should be used to base our decisions and risk assessments as leaders. We need to rise above *emotion-driven perceptions* and focus on the retention of our freedoms through *results-based action.*

In the meantime, you're welcome for those freedoms to attack me on social media and call for my resignation. That's your right to *opinion,* but not your *authority* to mandate...that belongs to a judge and jury. We should continue to encourage freedom of expression without censorship, but it doesn't mean I should resign based on your accusations, opinions, or hurt feelings.

Can't you see? That's the exact thing we're trying to defend, so don't bite the hand of liberty just because someone barked at you or did something to offend you. They didn't actually bite you. Your freedoms are still intact. So don't make it out that they did bite when there are no teeth marks, and you were simply offended. You're devaluing your speech when you falsely accuse, and leaders, you're leading scared when you base your decisions off such public opinions or pressures.

◆————————————◆

At the beginning of every morning during our Triple ACE training event, I rallied the troops in an auditorium and gave them a pep talk. On the last morning leading into the most pivotal day when all the Distinguished Visitors (DVs) would be in Cheyenne to witness the more complex operations of Triple ACE, I led with a quote I had put together that culminated everything we had accomplished and were attempting to finish that day. I had been researching "pioneers of firsts" in history and was stumbling upon the same common sentiments surrounding the Wright Brothers (first to flight) and Marie Curie (radiology chemist and first woman to win the Nobel Prize...first human to win it twice). Words such as "insatiable curiosity" and "love of truth" and "relentless resolve" kept coming up as common descriptors of these historic pioneers. These stories are what inspired the quote I put together for that final morning of Triple ACE:

"To be FIRST, you must possess the COURAGE to overcome fear and failure. Once this is achieved, the UNKNOWN can be uncovered and the impossible suddenly becomes REALITY."

We wanted to be the first ANG ATCS to convert to a CAOS, the first to break one thousand operations in three days, the first to hot pit refuel an aircraft in Cheyenne, the first to create a NEST (Non-permissive Environment Security Team), and the first to prove one squadron could spoke to six different locations and simultaneously control with a single TOC directing from the main base. Why? So that we could be the first to be called when our nation needed mobile air traffic services anywhere in the world, for any reason, at any time.

202

In order to be first at all of these things, like the quote states, we needed courage to overcome fear and failure. We were attempting to add so many new operations to our resume leading up to this event and doing so could've easily overwhelmed us to the point of giving up on some of the ideas we wanted to include, and instead only concentrating on a couple objectives. However, we had courage and trusted each other to carry the load collectively.

Once we gained this courage through our unity in purpose (even just the exhilarating thought of "what if we actually pulled this off?") began to make us believers in our own capabilities. We started gaining momentum toward all the goals we had set, and a positive flow of energy circulated the room during each planning meeting. We started to visualize these ideas as actually happening. We *visualized* our success. And guess what? All of those "firsts" became reality.

This, my friends, is the definition of relentless.

Relentless effort, relentless energy, relentless positivity all led to the success of Triple ACE. It has been the recipe we've been using in the 243d for the past three years. All that the Red-tailed Hawks needed was someone to empower them and believe in them. I also believe experiences such as Snowpocalypse, acquiring a $15M Deployable RAPCON for free, making it to Southern Lightning Strike on time by pulling a car behind a U-Haul, commissioning NEST from the dust, inventing Landing Zone Boxes, finding ways to get locally certified Landing and Drop Zone Controllers approved for exercises, and even making the morale fund stretch to purchase enough pizzas to reward the squadron, or hand carrying those promotion orders for signatures— all of these successful experiences of relentless efforts—caused Triple ACE to be hugely successful. It is a *culture and mindset* of relentless badassery that has now gotten us voted to be the first ANG ATCS to convert to a CAOS. It took many years of consistent, relentless effort from the Red-tailed Hawks, and permission to take ownership of their jobs. Lead relentlessly with your heart and gut, and your people will surprise you.

I plan on covering the "Rebellious" portion more in depth in the very next chapter entitled "Get In Trouble." However, I wanted rebellious to be mentioned as part of the other "R's" within this chapter so that the reader could understand the differences between them all. So, without providing the examples soon to come, I ask you to simply contemplate the famous quote from Steve Prefontaine, one of the most rebellious distance runners in history. He proclaimed, "To give anything less than your best, is to sacrifice the gift."

When I start explaining the "rebellious" things I've done during my career, you will realize I chose this quote to help illustrate that I was simply giving my best...and sometimes your best requires some rebellion. I don't believe in doing something half-cocked. What's the point of even attempting if your half effort or ill preparedness won't produce anything worthwhile? That's why I only take bets I know I can win. So when I decide to do something, *it is* going to happen. Some may call this "dying on a mountain." I call it "conquering a mountain," because who wants to become king of the mountain, and then die? That's dumb. So you have to be careful which "mountains" you want to rebelliously conquer. I don't give anything less than my best when it comes to leadership, and I won't sacrifice any "gift" I have by rolling over and giving up. Choose your battles wisely and you will win the overall war. More to come in the next chapter.

Staying on the topic of famous distance runners, I wanted to educate everyone on who Roger Bannister was and why he is still important.

Roger Bannister was the first human to ever break the four-minute mile. In May of 1954, he ran a 3:59.4 mile.

Just forty-six days later, his record was beat.

Before you brush past what I just wrote and think it's no big deal that his record was beat just forty-six days later, put this monumental feat in perspective. The first Olympic games were held in 776 BC, which was 2,730 years before Roger Bannister broke the four-minute mile. Therefore, for thousands of years, man had never run that distance that fast. According to trackandfieldnews.com there have been 1,663 humans to ever break the four-minute mile since Bannister first did, which was sixty-nine years ago, which means on average twenty-four new runners a year get added to that list (about two a month).

Cool, so what's the point? The point is that it took thousands of years for one person to show that it was humanly possible to run that distance that fast, and now in just sixty-nine years from that moment, hundreds have done the same thing. There is something to be said about Roger Bannister being considered a rogue athlete. Even though his record only lasted forty-six days, the barrier he broke by being the first, paved the mental belief path for 1,662 other runners. *That* is the definition of rogue.

I've already explained the many "firsts" that the 243d has accomplished. You can guarantee that this has now paved the path for the other ATCSs to do the same. I also hope that those reading this book will realize they can also "go rogue" and trailblaze within their respective professions.

Although we accomplished many "firsts," they were all improvements on what we already had been doing...more or less. The one item that is truly groundbreaking for our ATC enterprise is NEST (Non-permissive Environment Security Team). I definitely went rogue when I created this...without asking anyone's permission.

NEST has turned into something within the 243d that has given many something to be excited about again. More importantly, it has shifted the mentality of the entire squadron to become more combat-minded, regardless of whether you are a NEST Operator. If we are going to convert to a *Combat* Airfield Operations Squadron (CAOS), then we had better be prepared for combat, don't you think?

The Air Force has published tactics, techniques, and procedures outlining ATCSs taking handoffs from Special Tactics Teams, so shouldn't we be more tactical than the baseline Airman? If we truly are going to be placed on an island in the middle of the ocean and expected to defend ourselves for any amount of time while controlling airplanes in and out, shouldn't we give ourselves the best chance of returning home alive? To all these questions I say, hell fucking yeah!

So, I went rogue, and with the expertise of the prior Army paratroopers in my squadron, we developed a curriculum to meet these objectives. Will this turn into a nationally recognized program for the other ATCSs/CAOSs to mimic? This book is being published before we know the answer to that question, but you better believe for damn sure that the 243d will continue to train and produce NEST Operators who are prepared to seize, operate, control, and return (SOCR) the airfield. I owe it to my Hawks to best prepare them for the worst. Rogue moments in history sometimes generate "the standard" for others to then follow, just as Roger Bannister did.

◆————————◆

Relentless, rebellious, and rogue are character traits that are commonly associated with pioneering leaders. Careful application of these traits, mixed with many other balancing characteristics, can make a superior leader. However, one trait we should all avoid and be constantly self-monitoring for is recklessness. The line between the other three traits outlined in this chapter and recklessness can be very thin at times. That is why it's important to have trusted advisors and mentors in your life that can check your blind spots and give you a heads up before you get sideswiped.

Luckily, I haven't done anything reckless in my career. Nevertheless, I have come to recognize the feeling inside when I'm creeping closer to my limit and need to take a step back. It takes getting close to that line, to be able to recognize it.

One such instance that stands out in my mind, when someone else saw me bump up against that line before I did, was one specific morning wakeup with that same Director of Operations at OTS that became one of my mentors; the same one who told me to "Execute a prisoner and feed the troops" when I first took command. On this specific morning, he and I were running the same hallway, kicking down doors and melting faces in our usual fashion. Things were going fairly normal, but it was the second week of OTS and us Instructors had been doing morning wake-ups for about nine mornings by now. So we were all hitting the "wake-up wall." No excuse, but this did cause many of us to have shorter fuses when it came to how much we were willing to put up with chucklehead trainees and their repeat of Week 1 type of mistakes.

Once the initial wake-up shock-and-awe was over, I stood at one of my favorite corners to catch trainees not squaring their corners and yell "square yo corners!" Trainee after trainee would rush by, attempting to snap their way crisply around the corner. At this point in training, I would make the correction only when needed (instead of a general broadcast correction like the MTIs had taught me to use as an occasional tool to tighten everyone up). There was a lot of commotion and trainees making simple mistakes in this particular hallway, and trainees screwing up multiple times was like blood in the water for other Instructors to focus in on what else needed correction around that area.

I don't know what it was, but that particular morning the trainees' jar of fucks was nearly empty, because they were moving slowly and making simple mistakes we would've expected out of Day 1 trainees. As I was barking around my corner and other Instructors were swarming about as trainees bounced through the hallways like pinballs trying to exit the building as quickly as possible, I yelled at one particular trainee to square their corner...and under his breath I heard very faintly, "Whatever."

Obviously, this trainee's mental resiliency was beginning to break down, because talking back to an Instructor *is the last thing* you ever want to do in that type of training environment. I just about broke glass ordering him to report to me. You could see in his eyes the mystified glaze of regret immediately descending upon him as he stood at attention and reported to me. I'm not exactly sure what I said (I can neither confirm, nor deny) but none of it was sunshine, cupcakes, or unicorns.

Somewhere in the middle of seeing red, I educated this young trainee to recognize the importance of squaring corners and respecting rank. I also increased his vocabulary awareness by explaining that "Whatever" is only a word you say in the military when it's followed by "...is best for the Air Force, sir/ma'am." During this very educational experience for this young Airman, I could feel myself starting to slip from loud and controlled, to atomic and...less controlled. I saw my DO glance over at me with an expression of slight concern, so I quickly sent the trainee on his way to scurry back into the herd of kittens.

Later that day the DO asked if I had a second to talk in his office. As he shut the door, we started talking about how the morning wakeups went. He mentioned my little "educational session" with this particular trainee. He asked how I felt about that exchange. I explained that I felt like I let it get to me, instead of maintaining my usual controlled intensity. He was glad to hear I recognized this, because he spotted a difference in my approach as well. He just wanted to make sure I was doing all right and that if I hadn't recognized it, he wanted to point it out. Why? Because he cared about me and wanted to ensure I didn't need a break. Coming from a guy who rivaled me in intensity at OTS, it meant a lot. He wasn't being oversensitive or condemning my approach to the situation. As a mentor he just wanted to walk me through that moment to help me recognize that I was approaching my reckless line, so that I could ensure I never crossed it. I can't tell you how much I appreciated him taking me aside to memorialize that threshold.

Although my job no longer involves daily early morning wakeups and yelling at trainees, that same principle still applies to knowing my own limits and knowing when I am approaching a reckless line. I may have a higher risk threshold than most, but that should never justify moving my reckless line. That line must always be in check.

Another lesson this taught me was not to fight for authority I already had. As an Instructor, I was automatically in a position of authority over trainees. I didn't need to yell and correct to establish my position (nobody in any circumstance should). We yelled and corrected at OTS so that it created a controlled pressurized environment for the trainees to operate within. So outside of a military training environment, as the leader I should never feel the need to fight for relevancy or positioning. I've already got it by my duty position and/or rank alone. So don't give up your high ground to someone baiting you from below. It reminded me of a saying one of my childhood mentors would often tell me, "When you wrestle with a pig, you both just get dirty." Don't take the bait, don't enter a competition you've already won, and don't recklessly make decisions based upon the emotions of a taunt.

To end this chapter, I want to explain the *300* movie reference I threw out there in the last chapter. The movie is *very loosely* based upon the Battle of Thermopylae in ancient Greece. It would probably be better to say the movie was *inspired* by this historic battle, not based upon it, since the movie is far from accurate. Nevertheless, the true account is celebrated as an example of relentless persistence against seemingly impossible odds. King Leonidas demonstrated relentless, rebellious, and rogue leadership with his decision and strategy to bait the Persians' much larger army into a bottlenecked position against his much smaller army. This bold "funnel" strategy was working, until a Greek citizen betrayed Leonidas by informing the enemy of a path that took them around Thermopylae, eventually causing the demise of this strategy.

After suffering large casualties, the war council ordered a retreat of the majority of the Greek army. However, Leonidas and his three hundred bodyguards and about 1,100 Boeotians remained behind, since retreating would defy Spartan custom. This may be where many consider that Leonidas made a reckless decision, but this is the mountain Leonidas literally decided to die on. Whatever your stance on Leonidas's final decision not to retreat, you cannot deny that these warriors were relentless, rebellious, and rogue. They had purpose and understood the risks involved.

The most important part that I took from this movie, was the scene leading up to the battle when Leonidas was on the warpath, gathering allied troops to join him. When he comes upon the commander of one specific allied army, the commander comments disappointedly concerning how he expected there to be more soldiers in Leonidas's army. Leonidas proceeds to ask a few of the other army's soldiers what their professions are. The responses range from potter to sculpture to blacksmith. Leonidas then turns to his three hundred and yells in his war cry voice asking, "Spartans, what is your profession?!" in which in unison the three hundred grunt loudly "Ahuu! Ahuu! Ahuu!!!" while pumping their spears in the air. Leonidas then turns to the other commander and states, "See, old friend, I brought more soldiers than you did."

As a leader, how are you going to create a culture within your organization where your people have that type of dedication to their profession? What causes humans to become passionately dedicated to the work they do each day? I believe it starts with them believing in their own personal importance to the mission. They must find their own "why" for coming into work and doing their best work. As a leader, it's your job to help guide each of your employees to discover that for themselves. I can guarantee you that most people want to be part of something that makes a difference.

Leaders must be relentless in their pursuit for excellence within their organization. They must be a little rebellious at times to make the necessary changes to get their people what they need to move the ball down the field. Leaders can sometimes even go rogue and be a leader within an enterprise, showing other similar organizations what the future looks like, by taking them there. Be mindful not to be reckless and ruin all that work by dying on the wrong mountains, instead of conquering the correct ones.

Once you can turn to your people and they can look at their relentless, rebellious, and rogue leader with full commitment to charge the next hill, your organization will be unstoppable. It's time to take some risks, just as our forefathers did when they signed their names to The Declaration of Independence. Let's be badass and show the world our nation's strength through how we lead our people.

Chapter 15

Get In Trouble

I was a pretty good kid growing up and never really "got in trouble" by my teachers at school. The worst I can remember is when I was in second grade, I had made what I thought was a really cool contraption that served no purpose. I had taken a small block of wood the size of a deck of cards, hammered five nails through it so they were sticking out the other side, and then tied a thin rope around it about three feet long. I would swing it around and thought it was pretty cool. So yeah, I had definitely created a weapon...without thinking it was a weapon.

I took it to school to show one of my friends at recess. I started swinging my homemade medieval flail around, nearly missing my friend's head, and had gotten just a few good rotations in before of course one of the playground duties frantically starts running over with her arms stretched out, eyes bulging, screaming "Puuuut thaaaat DOWN!" She then proceeded to confiscate my contraption (weapon).

Long story short, the principal realized I had half a brain and no intention to do harm to anyone. He said I couldn't have my contraption (weapon) back and let me pick a toy out of his "treasure chest for good behavior" instead. My kids laugh their faces off when I tell them this story, but they also wonder how I wasn't expelled...because that's the world they live in now.

Point of this story is, any time I would come even close to "getting in trouble" in life, I would get the same pit in my stomach and large lump of guilt in my throat that I remember getting that day in the principal's office. I was conditioned to think that if I weren't perfect, or if I didn't meet someone's expectations, or if I strayed from the intended course, that I was in trouble. Nobody likes the feeling of being in trouble…or do they?

I started gaining more confidence as a teenager in sports and academics. In college I became more comfortable with not being perfect, but I still struggled with being "righteous." It really wasn't until I joined the Air Force, was forced way outside my comfort zone, and was surrounded by young men much "rougher" than myself, that I realized I was far from ever being in trouble. I also realized I was kind of a wimp, too much of a people pleaser, and still an unhealthy amount of perfectionist.

The biggest turning point for this part of my leadership development came when I had a spiritual awakening as to how I was raised, decided to take a different path, and at the same time went through the hardest period in my personal life. This all happened in the years leading up to and then during the time I was an OTS Instructor. I evolved into a much better version of myself. From here, my confidence in myself and who I was skyrocketed. A personal transformation occurred over decades that led me to realize my true potential, not apologize for pissing people off, and being okay with rebelling to the point of not being afraid of getting into a little trouble. I feel that by the time I was hired into squadron command, I had finally found myself and was operating confidently in my own skin.

Gaining the commander's badge and the authority that comes with that (G-series), was an injection of confidence that I'll speak to in a different chapter. Point is, I learned to trust my intuition and not apologize for loving my people and doing what I felt was best for them.

Since joining the Air Force, I have had many experiences of "getting into trouble." Ranging from being yelled at in boot camp for simply not holding onto the handrail while walking down the stairs, to being overtly berated during ATC training to help forge me under pressure. Most of these were "built-in stressors" to help me learn how to control situations under immense stress, and all of them were legal.

One specific instance of this was when I was about a week out from becoming fully certified as an Air Traffic Controller at Luke AFB, AZ, my first duty station. From the time an Airman enters boot camp to when they've passed ATC school and are finishing their on-the-job training ready to be certified, averages about eighteen months from start to finish. So being one week out from such a momentous milestone in my young career, was very exciting…and stressful.

My ATC Trainer was a hard ass. Very smart, almost to the point of demeaning. Nevertheless, he was good at what he did, and I was a stellar trainee. However, he knew how to push my buttons to get me to approach that line of breaking. As we were running a simulator problem on the radar scope this particular day, unbeknownst to me he had instructed the sim pilot to throw everything at me during this forty-five-minute simulated control scenario. He wanted one last objective met before putting me up for certification on live traffic: find my breaking point.

About thirty minutes into the sim, I was holding my own, keeping it all together, and feeling pretty proud of myself since it was the busiest sim I had ever controlled up to that point. Then with about fifteen minutes left, my trainer walks back to the sim piloting room while I'm still controlling, and discreetly tells the sim driver to throw it all at me. Suddenly, planes from out of nowhere start popping up, requests are made for multiple practice approaches by all inbound aircraft, adjacent facilities start calling and requesting point-outs on diverting aircraft into my airspace, and one of the F-16s declares an inflight emergency with a request to land immediately. As I was controlling, I started getting mentally discouraged thinking this was all super unrealistic and wasn't providing me any training value. I started to lose the picture. I was sweating, my hand was clinched around the radio headset transmitter, and my speech started to crack. I did not give up, I kept everyone safe, but it got ugly toward the end, and I was getting pissed.

The moment the sim ended, with my trainer sitting right next to me, I stood up, kicked my chair back and over, and threw my headset on the ground and said, "That was unrealistic bullshit!" To which my trainer (keep in mind he outranked me by two ranks) calmly said, "Well I guess you're not ready then, are you?" which only pissed me off more. The even higher-ranking Watch Supervisor was present in the room, and I yelled over to him, "Sir, this is bullshit!" at which time I had crossed the line. My trainer stood up, thumped his finger into my chest and said sternly, "Get your ass outside to the smoke pit and wait for me!"

Well damn, this was not good…but I didn't care, I was livid. I had spent the past eighteen months of my life dedicated to earning my ATC badge, and this guy thinks I'm not ready because of an unrealistic sim?! I paced back-and-forth outside in the hot Arizona sun, waiting for my trainer to come tell me how worthless I was and how I disrespected authority.

The moment I saw him exit the building and walk my way, I stood my ground and stared at him with fire in my eyes. He walked up to me with a pissed off face and stopped less than six inches from my nose. He looked me straight in the eyes, thumped me one more time in the chest with his finger and said, "Now I know you're ready."

Wait…what?!

My stern face quickly melted into one of question. He went on, "I have been pushing you for the past nine months, trying to get you to break, and you never did. I was worried I wouldn't find that breaking point during your training, and I didn't want you to find it during live control on your own, so I purposefully made that sim so unrealistic that it would make you have to break. Even then, you held your own during the sim, but it was the reaction afterwards when you kicked over the chair and yelled at me—that's when I knew you were ready to be an Air Traffic Controller. Now you know your breaking point. Never let yourself get there during live traffic. People's lives depend on it. Congratulations, you're ready. I'll submit the live evaluation paperwork request for your certification." Then he turned and walked back inside, leaving me to internalize what had just happened.

Not only did this moment teach me about where my stress limit breaking point was while controlling airplanes; it's also the moment I gained my swagger as a controller. Up until this point, I had to be submissive at boot camp, worthless at ATC school, and coachable during OJT. However, there is an evolution that must occur toward the end of everyone's training where they transform from submissive trainee to badass controller. We can't have timid controllers making life-and-death vectors and altitude decisions with a dozen airplanes under their control. They must be sure of themselves and their radio commands. They must have confidence in their abilities and be able to justify their choices.

I found this to be true as I later became an ATC OJT Trainer and then an ATC schoolhouse Instructor. Every trainee that made it successfully through the training (which on average is only around sixty percent of those who attempt) had this same evolution occur weeks before their certification.

This may be why controllers and pilots will compete over who is more badass, because both professions require that same swagger in order to maintain control of the chaos surrounding them. Kicking over that chair and talking back to my Sergeant goes against everything we are taught during BMT. However, this is what was needed in that moment for me to become the asset the U.S. military needed me to become. That is what they hired me for.

This started me on a trajectory of confidence that I had never before had in my life, even as a successful competitive athlete or even as a provider to my family (two other instances in my life that helped build my self-confidence and worth). I realized that sometimes it took bold actions to accomplish what I was hired to do. I learned to suppress that pit in my stomach and lump in my throat, because sometimes pushing the envelope is the right thing to do. This is what I mean by "get in trouble."

———————————◆———————————

I do not intend to write about every moment in my career when I pushed the envelope or fought back. For one, it would make this chapter very long. For another, I don't intend to point fingers and call certain people or organizations out that could hinder the furthering of good relationships. In general, honey will get more done than vinegar. However, you can't just rollover at the first sign of resistance, nor should relationships between different inter-agencies be built upon soft feedback. Sometimes hard talks must occur. I've shut the door on the office of someone outside my organization before and chewed their ass. Sometimes it's necessary, and especially in the military we can't afford to be soft on each other. It goes both ways and I appreciate straight-shooting feedback on the chin.

I will share some instances and glaze over many others. The intent of this chapter is to give the reader permission to take risks, respectfully question authority, and be confident stepping into the unknown. An additional purpose is to remind those with the proper authority that it's part of their job to leverage that authority on behalf of your people. I've made many calls to other organizations when they weren't treating my people with logical professionalism. Sometimes you must call people out and use your rank to help those with less rank. You're a family, and you don't let anyone take advantage of any member of that family.

With this in mind, I'm not calling one particular HRO out, but in general I have had the most "rub" with these kinds of offices. Understandably, these types of organizations are founded upon very rigid regulations and it's their job to ensure adherence to labor laws, hiring practices, union charters, etc. Nevertheless, working within a system (ATC) that has its own very rigid regulations, I also know there

217

are plenty of gray areas, waivers, and loopholes in every organization that can help apply logic in order to do the right thing by the people.

Whether it's finding a way to qualify an applicant after an advertisement closed out because they accidentally answered a confusing prescreening question incorrectly (but they are clearly qualified to board for the job), or not making someone retake a three-day supervisor training course because a new software didn't migrate everyone's previous supervisor training certificates over...you need to find a way to apply common sense and make it right. Don't get so caught up in the rules that you forget about the people. Without going into details of this area, I just encourage you to find that balance with inter-agencies where mutual respect promotes helping each other out. Nevertheless, as JFK once said, "Let us never negotiate out of fear. But, let us never fear to negotiate." Help your people out by not being afraid to negotiate with other organizations that are operating out of fear.

Another area I don't want to dabble too deeply in because of its confidentiality is the legal representation your organization may have. In the military it's the Judge Advocate General's Corps (JAGs), aka: attorneys. As a military G-series commander, I have direct access to the JAG. The JAG is not a fellow commander and never will be. One of their primary duties involves providing legal counsel and advice to commanders. It is extremely important to have a great relationship and mutual understanding with your JAG, or other legal representation if you lead a civilian organization.

My first "serious" incident that required me to seek counsel from my JAG came fairly early in my command. Up until then I had sought basic counsel on a few discharges and union disputes. However, when one of my controllers very suspiciously tried to dodge a random urinalysis drug test, and then proceeded to throw up in the waiting room of the drug test facility on base, I immediately ran over to the facility and called the JAG to meet me there. The report from the escort that had accompanied him to the restroom to provide a sample (at which time the Airman made excuses for not being ready to provide a sample), was that he suspected he was hungover. This was my first instance where

Security Forces (military police) may need to get involved. As a commander, I have certain authorities to order drug tests, breathalyzers, etc.

I counseled with my JAG concerning my options. This JAG advised against breathalyzing the Airman, they didn't think there was enough probable cause. I knew the escort (he was one of our Hawks as well) and trusted his judgment. I overheard that the Airman had been drinking the night before at someone else's house, so I stepped aside and called the Hawk that had hosted the gathering. They confirmed that this Airman had been drinking, but that he didn't break the twelve-hour rule (controllers can't drink within twelve hours of their duty day). I also had the fact that he threw up, and since this was during COVID, I felt I had enough probable cause needed to order a breathalyzer and COVID test. The JAG once again strongly advised against this; they thought it could come back to be an overreach of authority without enough cause. I respectfully thanked the JAG for their advisement, and then proceeded to call Security Forces.

Without sharing confidential details, a breathalyzer, special investigator interview, and sobriety test were all conducted as I had ordered. In the end, I had the evidence needed to discharge this Airman for showing up to work under the influence (I also didn't allow him to leave the building that day until he provided the urine sample for the drug test).

The miracle of this story is the fact that this Airman was randomly selected for a drug test that day, which prevented him from going to the tower to control live planes, since he had to go provide a urine sample first. Then the rest of the story ensued, so he never made it to the tower. If he hadn't been informed at the beginning of his shift that he had to go drug test, he would've controlled live traffic under the effects of alcohol. Truly a miracle that things played out the way they did.

He and I later had a heart-to-heart talk when I read him his discharge paperwork. He understood the lifechanging consequences he had made that day, which could've ultimately led to lives being lost. Nevertheless, this story has a happy ending. As I mentioned in a previous chapter, I still check in with those that I kick out. This particular Airman used this pivotal moment in his life to change things around completely. He found a spiritual pillar in life, moved out on his own away from the house of younger men not making good life choices, got a better job, and has a girlfriend that is a positive influence in his life. Although this isn't the reason for sharing this particular story, it cannot go without mentioning.

The leadership lesson from this story is that as a young commander with little experience in the seat, I still went with my gut—which ended up being against the council of the JAG (who by the way also outranked me). Don't let the fear of "getting in trouble" fog your common sense judgment.

•◆————————◆•

Getting called into your boss's office to be told to slow down isn't a bad thing. I can't tell you how many times in my career I've been told by my boss or supervisor, "HH, it's a marathon, not a sprint." To which I reply, "Yeah, but I sprint marathons, sir/ma'am." They always shake their head and smile, knowing they can't truly put me on a leash. They always appreciate the energy and commitment and are grateful that they aren't having to ask me to step up and start working harder.

Nevertheless, this doesn't mean I haven't over-sped before, and been ordered (not asked) to slow down. I also appreciate these moments. When you are coachable, moments like these are indicators that you're doing the right thing. I say this because I occasionally have to pull one of my own Hawks aside and carefully communicate that I love their thrust, we just need to adjust the vector slightly. I never want to diminish someone's passion by telling them to slow down or stop being so excited. Like I've mentioned before, I've been lucky enough to have good bosses (mostly) who are appreciative of my style and learn how to mentor me through times of over-speeding without putting me on a leash.

One such instance was about nine months out from Triple ACE. We had only just conceived the larger concept of what we wanted it to look like. From conception of idea to execution of event, was only ten months total…which is an extremely short timeline for a training event of the magnitude that Triple ACE turned out to be. So, after a month of ironing out some details and developing my slide show for selling the idea to other units, I first had to sell the idea to my Wing Commander. I did this at a meeting where all the other commanders on base attended, along with some other leaders. I threw my slides up on the screen and worked the room with my Concept of Operations (CONOP) brief. I was energetic, quick, and groundbreaking.

Once I finished, I could see a little concern on the Wing Commander's face. He asked a few questions, mainly focused on the part about landing a U-28 spy plane on a highway in the middle of Wyoming. He asked what coordination I had already accomplished for that piece. I instructed him that I had a meeting already scheduled with the Director of Wyoming Homeland Security the very next day to go over the plan. His jaw dropped and he asked how I got a meeting with the Director. I explained that one of my DSG Hawks worked in her office and that's how I got her number. Another jaw drop. Then silence. Then, "Okay, HH, thanks for the presentation. Let's talk afterwards about a few details."

After the meeting ended everyone exited the room except the Wing Commander and his Vice (who used to be my boss when I first took command, so he knew me very well). They asked me to sit across the table from them. The Wing Commander was a much more reserved man, and not one to raise his voice or come at you head on. However, the Vice (my old boss), although still very calm and collected all of the time, knew how to put it right on the chin when needed. It was the Vice that started the conversation. In fact, it wasn't a conversation, and that's exactly how he started, "HH, this is not a two-way conversation and I need you to take a moment to open your mind to some heavy mentoring right now."

Me: "Yes, sir." As I pulled out my notebook and pen.

Vice: "HH, I am no longer a G-series commander and technically can't order you, but I am going to strongly recommend to the Wing Commander (as the Wing Commander nods his head in agreement) that you will not be attending that meeting you scheduled with the Director of Wyoming Homeland Security tomorrow. What you have proposed regarding the highway landing is not going to happen and you will not push for that piece of Triple ACE any further. Additionally, I am going to show you on this piece of paper how things like this work, and you will see afterwards nowhere on this paper where a squadron commander goes directly to the Director of Homeland Security for anything, much less a highway landing in the state. Colonel (as he turns to the Wing Commander), what are your thoughts?"

The Wing Commander then ordered me exactly as was outlined by the Vice.

Me: "Yes, sir."

The Vice then went on to explain how coordinating something like this works. It was not a demeaning session, it was actually very educational, and you better believe I took good notes as he was talking. He explained how this same thing was attempted years ago by an out-of-state unit and it failed horribly due to poor relationship building and planning. Come to find out, there were *a lot* of organizations needing to be involved with this type of operation, and planning needed to start at a minimum a year out. He went on to explain that going straight to the Director would've caused the scars of the previous attempt to be ripped back open, and that everyone up to the TAG (two-star General) and Governor would be very upset. He ended by telling me how excited he was about the rest of what we had planned and supported all other aspects of Triple ACE that I had presented.

Once he was finished, he said, "HH, now you can speak. What questions do you have?"

Me: "Sir, thank you for taking the time to pull me back from a train I didn't see coming. I understand now the implications of my proposed plan. I will definitely cancel tomorrow's meeting and proceed as ordered..." (they both nod their heads in approval while I pause for dramatic effect), "...but, now that I know the process, have the notes needed, and have both of you here, I would like to propose for your approval a highway landing one year from now."

They both grinned, shook their heads, and chuckled under their breath. I smiled back and shook both their hands as we left the meeting. Gotta keep 'em on their toes, right?

I know I probably give my Group and Wing leadership multiple heart attacks a month, but I truly appreciate their perspective and mentorship. Luckily, I have that type of relationship with them that allows for give-and-take. Point is, I "got in trouble" by two full bird Colonels and didn't let that defeat me or prevent me from adjusting the concept of operations to still make Triple ACE the success that it was. Even if I never get an aircraft landed on a highway (I will), I was at least attempting to push that envelope past the usual acceptable tolerance to see what was possible. Guess what I found? A whole lot of mentorship and education. Now I have that knowledge, and my bosses also continue to evolve to support and coach my style of leadership.

◆————————◆

Unfortunately, we have allowed our minds to get "soft" with the conveniences of the modern day. Naturally, our mental tenacity has deteriorated with every new invention that takes the struggle out of daily life. This is why we have to pay monthly memberships to gyms in order to have a place to attempt to replicate that basic need for mental strengthening through physical exertion, because we don't get it naturally through our daily lives (most of us don't). I'm not saying I'd like to go back a hundred years and surrender my air conditioning and car. I'm just attempting to draw a parallel between how we used to challenge ourselves mentally through physical means, and how the need still exists to actively build mental resiliency.

In the movie *The Last Samurai*, the main character Nathan Algren (played by Tom Cruise) is a U.S. Civil War Captain that has been hired by the Emperor of Japan to teach their army regarding modern warfare. Since the Americans had come to perfect the current firearm and gatling gun during battle, the Japanese wanted to leverage these more effective ways in their attempt to wipe out the Samurai. When Captain Algren is captured by the Samurai, he eventually learns their ways and gains mutual respect; recognizing that their warrior spirit greatly outweighed that of the modern Japanese soldier. Captain Algren is later given a samurai sword by his captor with the engraved words, "I belong to the warrior in whom the old ways have joined the new."

While I was recently rewatching this movie, I couldn't help but think about how this applied to our current warfighters (and private sector "business warriors"). The first battle in the movie between the modernized Japanese fighters and the Samurai warriors showed the Samurai winning the battle with fierce dedication to their craft and each other. Conversely, the Japanese fighters struggled to even shoot their modern weapons at the fierce warriors charging their way.

Why have we given up on the warrior mindset? Is it due to the ease of how we now battle? Have drones and satellites replaced the warrior? If we aren't being challenged with kinetic war face-to-face like we did for so many centuries, will this soften our minds over time?

This is why I still try to join "the old ways with the new." I believe in pushing my people to their limits in all aspects of their life, and sometimes this is controversial.

At OTS, this was easier to implement because we had physical means to teach mental lessons. A couple times during the course we would run in formation as a flight, in competition with the other flights. Envision two columns of sixteen Airmen running in unison. Now multiply that by about eighteen flights, in the dark Alabama morning, singing jodies, running down the streets of Maxwell AFB. It was always such an inspiring moment to hear and see nearly three hundred Airmen running around the base chanting, carrying their banners, training to become military officers.

However, it was a competition, and I don't believe in catering to the lowest common denominator.

The rule was your time didn't stop until every member of your flight crossed the course's finish line. We also have a saying in the military, to "never leave an Airman behind." Most Instructors took this to mean don't leave one of your trainees behind on the course who is sucking wind and dragging ass. I took this to mean, don't leave one of your fellow Airmen lying there to die when they took a bullet on the battlefield.

Bro, we're on a fenced-in and guarded military installation. If one of my trainees starts falling behind, they can find their way to the finish line. If they end up having some sort of medical emergency, there are hundreds of other trainees running staggered start times on the same course, and a medical truck driving that course over and over during the run to ensure nobody is hurt. So no, I'm not going to slow the entire flight down so that you can catch your breath or make an excuse for why you can't maintain a steady jog for five miles during a military accessions training run. I don't buy it. And that's exactly what I would tell my flight before every run, every class. "We will not slow down, you will run at my pace, and if you can't, you will find your way to the finish line as quickly as possible."

Guess what? Although I would occasionally have an Airman fall back, they always kept the rest of the flight in sight as they pushed to keep up, and we ended up winning…a lot. In fact, one class we swept all four physical fitness challenges. Your body is capable of amazing things, if you don't let your mind give up before your legs.

I also didn't believe in cheering in the last person walking their ass across the finish line. Why do we do this? What are we cheering exactly? "Good job on being last and not coming to OTS mentally and physically prepared to become a military officer?" No. Let them feel the pressure of letting their fellow Airmen down by not holding up their end of the deal. Does this sound harsh? Yes, in today's world it does. But do the other ninety-nine percent we defend want to see their "elite one percent" walking across the finish line of a five-mile jog acting like they just climbed Mount Everest? Pretty sure they don't. Pretty sure this says more about someone's mental tenacity than it does about their legs and lungs.

So yeah, I may take the quote on the samurai sword to heart when I see modern war tactics needing to be joined with the old school warrior mentality. The need for mentally tough warriors remains the same, whether you're swinging a sword or clicking a computer mouse to kill the enemy.

The reason I include this story in this chapter is because I would "get in trouble" (kind of) for doing this. Not by the commander, but by fellow Instructors and the supervisor of the Instructors. They would say I was putting winning above safety, or that I was hurting the feelings of those that were slower runners (that one was my favorite). Since we were all the same rank and they weren't my commander, I told them I thought they were creating weak-minded officers, and then thanked them for the trophy when we won the events.

At the end of every OTS class the night before graduation, we all dress up for a fancy catered dinner in our mess dress (military tuxedo). This Air Force tradition is called "Dining In." I remember the class where we swept all four events, the student emceeing the event (who is supposed to be funny and crack these types of jokes) threw this line in his commencement speech, "...and we will never leave an Airman behind...unless you're Captain Hochhalter." The place erupted in laughter, and I, too, couldn't stop laughing at the crack. Well played.

When I sent our Triple ACE video out to anyone and everyone, I received great feedback and a positive reception to our operations. However, the part where it shows our NEST Operators executing seizure of the control tower by infilling off Blackhawks with their faces painted and weapons at the ready, I received a text from our WSC Chairman at the time (a different WSC Chairman than the other two I already mentioned in a previous story). In short, he was doubting the need for NEST. This was my reply via text (with acronyms spelled out for the reader):

"The Combat Airfield Ops AFTTP [Air Force Techniques, Tactics, and Procedures] shows ATCS overlapping with STT [Special Tactics Teams]. Which means we are supposed to take the handoff from STT after they have seized an airfield. NEST [Non-permissive Environment Security Team] bridges the gap between STT and ATCS/CAOS so that this handoff occurs smoothly AND so that those STTs can forward deploy to another airfield and do the same thing, instead of staying behind to provide security to us because Air Force doesn't know basic security tactics. If STT can leave us behind and know we're in safe hands by providing our own security, then they can open another airfield and another CAOS can come in behind them and start controlling that next airfield. And yes, it's an MCA [Multi Capable Airman] initiative that will eventually become an SEI [Special Experience Identifier] so that any CAOS AFSC [Air Force Specialty Code, aka: career field] can become a multi capable airman. It's already brought me five new recruits and helped me retain six others. If we're going to have "Combat" in our name, we had better know how to be combat airmen."

He replied with some other jabs, but the one that got me 'Fired Up!' was when he texted: "I sure as hell didn't become a 13M [ATC officer] to go back to clearing rooms. ...most of these cats are controllers and maintenance technicians, not warriors...nerds bro, true story."

To which I replied (keep in mind, this is an O-5, but a fellow commander): "You can question our training approach, and you can even question the need for NEST, but I sure as hell am not showing up to some island in the Pacific unprepared. These are my Hawks and I owe it to them to get them as prepared as possible for the next war. I'm saddened that I'm going to be fighting you for support on this one, but it hasn't stopped me in the past from accelerating past naysayers. I didn't become a 13M to not be prepared to use my weapon. And your guys may be nerds, but mine are warriors!"

So, you can see how over time I've built a resiliency to that pit in my stomach when I "get in trouble" and why I believe that as long as you're not being reckless, it's a good thing to push the envelope. This is after all, what helped me grab the TAG's attention with a white paper about needing a new radar, got us coined by both the three-star DANG and four-star CNGB for the successes of Triple ACE, pushed our squadron to Fort Carson on time during Snowpocalypse, obtained multiple meetings with the Mayor of Cheyenne for public support, put the squadron on the map at the rodeo parades, gave me the gumption to tell a fellow commander that I was indeed bringing my DO to the WSC and another commander that he was wrong about NEST, convinced our logistics squadron to issue us our weapons for the first time ever, cold-called the F-16 squadron commander to get them to commit to Triple ACE, demanded at a table with nine Lieutenant Colonel commanders that we not delay the vote on the conversion to CAOS, held a lumberjack-themed holiday party on base with a homemade ax throwing range in the warehouse, and how my SEL and I were able to confidently walk into the offices of every high-level leader in the Air Force at The Pentagon to hand them our squadron Christmas card...just to name a few.

Leaders, be bold! Be strong! Combine the strength of modern technology with the proven need for mental tenacity, to regain the warrior ethos once again! So long as you are not reckless, go ahead and get in trouble, and show your people that you are relentless, rebellious, and rogue...because you care for them. Always Fired Up!

Chapter 16

Your Authentic Recipe

You've seen me write 'Fired Up!' quite a few times throughout this book. If you work with me, you'll likely hear me yell it half a dozen times a day. It's my war cry, the catchphrase I use to end my Commander's Calls, and it's on my commander's coin and email closing salutation. It's just what naturally started coming out anytime I got really pumped about something...which is why I say it a lot. I will sometimes randomly shout it out in the middle of a meeting when I agree with what's being said (much like "Amen!" during a sermon in the south). I'm pretty high energy and genuinely love what I do. So it's a natural thing to hear me blasting my music down the hallway in my office, occasionally clapping to the beat, and frequently yelling out, "FIRED UP!" when I get good news. I can't help it, it's just who I am, and I'm not ashamed of it.

When I went through OTS as a trainee and then again when I instructed there, you could pick out the Instructors that were forcing it with their yelling, instruction, and demeanor. You could see right through the ones that were just trying to mimic what they saw some other Instructor do that they thought was effective. They weren't ever fooling anyone, so I appreciated when Instructors would remain true to themselves.

I met one such unique Instructor when I first arrived at OTS for Instructor duty. Since we both arrived at the same time, we were in initial Instructor upgrade together, and to this day he is probably my best friend in the Air Force. Our classrooms in which we taught ended up being right next to each other, so we shared a wall. He is one of my favorite people in this world. We'll call him Captain Zen, because he would have his trainees do things like meditation exercises…he had a very non-standard approach to teaching.

The two of us couldn't have been more opposite when it came to instruction style. For example, he was conducting one of these "peaceful mind exercises" during the "Conflict Resolution" lesson by having his trainees sit cross-legged on the floor and close their eyes (hippie, dippie, bologna), while I was next door in my classroom teaching the same lesson, but slamming things against the wall so hard that it caused a frame on his wall to fall and break. We both were genuine to ourselves, and we both produced stellar officers. We just approached the same subjects differently. We also laugh harder than ever when we're together. We joke that we're twins separated at birth (which makes people chuckle, because he's black and I'm white).

I spent my entire tour at OTS teaching alongside Captain Zen. Also, my Air Advisor deployment coincidentally overlapped a couple of months with his deployment to the same country (completely different career fields and also different bases within the same country, but you better believe we found plenty of occasions to meet up). He has also visited me and my family in our home in Colorado, as I have returned to visit him. So, I can truly attest that how he taught and acted at OTS is who he really is. He wasn't putting on a show or trying to be someone he wasn't authentically. He is able to reach people differently than I can, and I respect the hell out of his approach to leadership. He truly "gets it" and makes a huge impact on the people he works with at every duty station he has been assigned. I can also attest that he is the same person in the uniform, as he is outside of the uniform. This is authenticity.

How you do anything, is how you do everything.

I have outlined leadership tactics, provided gouge, and shared "war stories" throughout this book. I have written strongly and under the premise that what I do works. However, I am far from perfect and make many mistakes; ask my command staff (and my wife). The things I have written are to assist in finding your own authentic recipe for success as a leader (or future leader) who is attempting to operate at the next level. Be true to who you are, but if you aren't sure who you really are yet, or are in the middle of an evolution, great! As leaders we must always be open-minded to adjusting our ways to fit the current circumstance, problem, or era.

What worked for me at OTS, didn't completely translate over to the 243d. What works for me now with the 243d, likely won't transfer perfectly over to whatever my next position is within the Air Force (granted I don't get fired before then). However, one thing I can guarantee you is that my actions will not be hypocritical to who I am at the core. This is why every person must constantly conduct self-assessments in order to stay true to themselves. Otherwise, people will see straight through your attempts at being someone you aren't. We see enough of that on social media and we have all gotten pretty good at sniffing out the bullshit. Make sure how you do anything, is how you do everything.

•◆——————————————————◆•

Something I've discovered over years of physically working out and trying different routines is that my body has a natural "normal" it's constantly trying to maintain. I'm not a big or small person, I'm about as average as they come. I can usually pull anything off the department store clothes rack, and it fits how the mannequin shows. I'm also thirty-seven, so I ain't growing any taller. As much as sometimes I wish I were a 6' 2" 230-pound linebacker, there is just no way I can get there. I may be able to get to the weight and muscle mass, but even then, it wouldn't be sustainable. My body wants to be a certain "type" and over the years

I've perfected what keeps me feeling my fittest. I've done the same thing with what I eat. I've found how my body reacts to certain things and found what keeps me healthy and naturally energized. I have also come to accept my imperfections and own what I *do* have that makes me feel confident in my own skin. I'm not in love with myself by any means, but I don't judge myself for things I can't change either. I just confidently own what I have.

Occasionally, I'll have a co-worker come ask me what I do for workouts. They comment on something they want to be that they currently aren't, and they think that if they just followed my regimen that they'd have the same results. I can't tell you how many times I've willingly written down my workouts and also added how I eat (because diet is the larger contributor in my personal experience), thinking that person will actually do it and find the same results. More often than not, they simply don't follow the plan consistently or long enough to see results. We're talking they give up in less than thirty days. No way you're going to see true results in that short time. Secondly, if they do stick with it and start seeing results, they find their bodies aren't changing to look exactly like mine. Go figure! We're biologically different in so many ways!

After seeing this over many years of providing recommendations when asked, I stopped giving details on what I do that works. Instead, I start asking questions about what they currently do and what they've tried in the past. I then sprinkle in some recommended workouts and eating behaviors that might fit their body type and lifestyle a bit better. However, I don't spend a lot of time on this. I mainly focus on hammering the point that they need to listen to their body and give it what it's asking for.

I deployed with a very tall and much more muscular person than me. For some reason he envied my lighter, more agile running abilities. He told me he wanted to become a runner. I recommended that he didn't. Instead, I recommended he try rowing, swimming laps, or spin classes to get the cardio he was seeking. Running would've been too hard on his joints and his body naturally wanted to be bigger. This is why I recommend always finding what works for you; things that will be somewhat enjoyable (not that every workout is going to be super fun, but it shouldn't always be dreaded either) which can be *consistently maintained*.

I share all of this to help guide your same process for finding what works for you as a leader. To find your authentic recipe. It's got to come naturally. You shouldn't be fighting yourself to become a stellar leader. If you do find after multiple serious efforts with different styles that you just can't find the passion in leading, then maybe leading isn't for you. I'm not saying that to be degrading or frame it as a failure. Truly. Some people (most) just aren't cut out to be good leaders, and we all know more bad leaders than good ones, so don't force it. Maybe you'd be better as a manager? Maybe you're the best at what you do skill-wise, and it doesn't require you to be a leader or a manager. That's awesome! Please, continue to be a hugely appreciated asset to the team! Remember, there is no leadership without a team to lead.

We need those who are passionate and great at what they do. This is why I think the Air Force and Space Force should adopt Warrant Officers, like the other military branches. Warrant Officers rank between the commissioned officers and the enlisted. Every Warrant Officer I have met absolutely loves their job because they get to do whatever their job is, for their entire career! They don't have to worry about being put in command or becoming a superintendent. It's the perfect fit for many professionals. An example of this are Army helicopter pilots. Most (not all) Army helicopter pilots are Warrant Officers, and they just love flying the bird and being the best at that mission set. The commissioned officer pilots get put in charge of the warrant officer pilots, and the warrant officers are grateful. So, there are places in the military and in the civilian sector for those who don't want to lead or manage, and we need them!

Nevertheless, if you've gotten this far reading this book, it's likely you want to lead. So, I won't focus on this aspect any longer. I just wanted to add insight that as leaders we need to know how to guide our people in the direction that will best suit their strengths and natural abilities.

We want everyone in the organization to love what they do and know that leadership will help them reach their potential.

◆———————————◆

Driving back down the road of authenticity, I want to share some of the other ingredients in my own authentic recipe.

My office. Whether it was a desk in an open bay, secluded cubicle, or the current large commander's office I have the privilege to temporarily occupy at the moment, I have always taken ownership of that space. I love when people's personalities come out in their choice of how they decorate their offices. As previously mentioned, I'm lucky to have a wife that is naturally talented in the area of decorating, so I get a lot of help from her to make my spaces stand out. She knows me better than anyone else, and I've also got a lot of opinions when it comes to decorating my office, so it always fits my personality. It is very stately and traditional, with paintings such as the one of George Washington, dark wood furniture, and leather seating accommodations. Whenever I bring someone new into my office, they immediately comment on the presence it brings. It helps them understand my style of leadership.

A different approach was taken by my previous DO, who went to the local thrift shop with his teenage daughters and bought funny old grandma-looking art and ceramic pieces to hang on his walls, next to his Star Wars poster and Bob Ross lunchpail. It was intended to be funny, and it was. It totally fit his personality and he was being authentic to himself. Both offices spoke volumes and provided conversation pieces.

My speaker. I've always loved all types of music; I was raised on it all. I have musical parents, with my dad having played jazz trombone and my mom the concert violin (in high school my dad also played football and my mom was a cheerleader, so quite the variety). I was raised playing the violin, piano, clarinet, bass clarinet, and tenor saxophone…and just like German class, I don't remember how to play any of them now. Nevertheless, I found an appreciation for all types of music, and sports. As I've stated before, if you put my music on shuffle, you'd be on a historical journey over the past half century's worth of music. So when I took ownership of my classroom at OTS, I decorated it similar to my office style, but also was sure to buy the biggest portable speaker I could carry around. It became one of my trademarks at OTS, for better or worse (depending on who you were). Not everyone loved my choice of music, nor did everyone appreciate when I would blare it as I walked down the hallway to my classroom to start a lesson (see the previous chapter "Get in Trouble"). I owned it and my trainees loved it.

When I became a commander, I bought an even bigger and louder speaker. My first Commander's Call I walked in blaring Eminem as they called the room to attention, and the music hasn't stopped for the past three years.

Another ingredient in my recipe is humor. I love laughing, and most of the time it's at myself. I am lucky to have built a strong enough relationship with my people that shots fired can also be returned. Additionally, I like playing workplace pranks and showing my people the less serious side of me. I let my natural passion for life and my job come out. So when I had to be TDY for six weeks to a professional military education course in Alabama while a commander, I got with our Public Affairs office on base and together with my DO, we made a hilarious five-minute video. Most of it played out in the weightroom with me straining to lift light weights while I spoke about the twenty-five-year anniversary of the 243d ATCS being celebrated while I was away. We would cut back and forth from weightroom to random places on base, trying to turn everything I talked about into the number twenty-five. I also had a huge metal art of our squadron logo made to mount on the wall, with the twenty-five-year anniversary dates underneath.

After the video and wall art were shown at drill while I was TDY, the messages flooded in commenting on how funny it was and how they appreciated hearing from their commander, even while he was TDY. I did the same thing while I was deployed as an Air Advisor in 2023. I made a couple videos of me speaking to the camera at different locations, and then compiled nearly 150 photos of the squadron over the past three years and had Public Affairs put them to some of my favorite songs. It was a great way to remind the Hawks of their hard work and accomplishments over the past few years, while also staying in touch by telling them how much I missed them.

Helping to create ownership within others is essential for any organization to thrive. When others besides yourself start picking up little, almost unnoticeable, pieces of trash off the floor in your building and parking lot, you know you're making progress. When I first arrived at the 243d, I wanted to feel like I was part of the brand and was looking forward to wearing the squadron patch with pride. However, I quickly realized that the 243d didn't really have a brand. Although they had a traditional squadron patch, there was not a lot of squadron gear (t-shirts, stickers, etc.) and no squadron mascot. Not to get cheesy, but mascots help create an identity; look at any sports team. It doesn't matter how ridiculous the mascot, even Texas Christian University makes a horned frog look fierce, and their fans rally around it.

So I made it my goal within the first year of command to create a squadron mascot and morale patch. About six months in, I sent a survey out to the squadron to submit ideas for a mascot, and then a follow-up survey to vote. The one that had the majority of votes was Red-tailed Hawk. I was super pumped. The Red-tailed Hawk is a bird of prey native to Wyoming, is rustic in color and features, and one can often be found perched upon the railing that surrounds the catwalk outside the control tower. I sketched an idea for the logo onto a sticky note, took a picture of it, and sent it to my brother-in-law who is super talented with graphic design. Within a week, I had a professional logo. We then got to work designing t-shirts, koozies, lanyards, patches, pens, stickers, even temporary tattoos! I wrote up the significance behind the mascot and gave meaning to the different parts of the hawk and patch. We had the big mascot reveal at a Commander's Call and then opened up the swag store for all the gear available to purchase.

The Hawks swarmed to the table to purchase their gear! It was awesome to watch as later that day people walked around proudly displaying their swag. You could see the excitement of ownership growing. We also renamed our auditorium "The Hawks' Nest," our conference room "The Red-tail Conference Room," and our monthly newsletter "The Hawkeye." Word around base started to spread. It helped that we sent the Hawks on a mission to "zap" (stick squadron stickers) anywhere and everywhere they could around base. People from other squadrons would lift their coffee mug to take a drink, and the people around them would see our sticker zapped to the bottom of their mug. We put them everywhere. It became a common rally point for us to identify with, and it has taken off ever since.

Now each year we have a competition for designing that year's morale t-shirt to be worn under our uniform, and I get pictures from fellow Hawks all the time showing where they last zapped our sticker. We even have Hawks that stop on the freeway anytime they're on a road trip and pass a 243-mile marker and zap the sign. I also get text pictures of the Red-tailed Hawk anytime it lands on the control tower catwalk. He has come to be known as "Tony." I love getting pictures of mile markers zapped and "Tony Hawk" perched on the catwalk. It confirms that ownership is taking hold. It's a starting point for deeper ownership and it gives us a sense of collective belonging.

We've also painted the walls in our squadron building, hung camo netting, and had our logo printed onto large four-foot-wide wall stickers and plastered the hallways with those. We've also added vinyl lettering to the walls, framed the mission and vision statements, and proudly display our annual squadron group photos down the command section hallway. Also, you already know about the lumberjack holiday party "family photo" we took with all of us in flannel, which made its way to The Pentagon. We have fun here.

All of this is part of the recipe I've used in the 243d Air Traffic Control Squadron in an effort to build ownership, comradery, and drive. It's my style and although it might not reach every person within my organization, it does provide permission for those I don't reach, to find their own "why" behind coming into work each day, and also gives them permission to make it their own.

———◆———

Why are you here? What is your purpose? What is your potential? You've read these three questions multiple times now throughout this book. My authentic recipe is the culmination of a big bag of different leadership tactics, but they all boil down to being able to answer these three questions. If your people can't answer these questions with positivity and progressive actions after you've settled on your own recipe, then maybe you need to rethink your ingredients.

Also remember, not everyone is meant to lead. Not to disrespect Abraham Lincoln, but I disagree with one of his famous quotes. He is credited with penning, "Whatever you are, be a good one." We all know what he's trying to say here, and I'm only bugged by the word "good." If I may, I would like to switch that one word out for "excellent." I think doing so helps drive my point home concerning where you are and where you want to be. Whatever it is you want to be, just be sure to be excellent at it! Doing so might lead you to other paths you never thought you'd walk down. It will also give you great satisfaction in your daily work. Good, better, excellent.

Whatever it is, leadership or something else, discover who you genuinely are and own it! Be proud of whichever recipe it is you find that works, stay open-minded to the need to evolve with the times, and then press forward with joy. There is too much in this world to be sad about; don't let your work be one of them. Keep working on your own authentic recipe until it is excellent...and then keep working to keep it relevant to meet the constantly evolving needs of your people. Be an authentic transformational leader.

Chapter 17

No Secret to Losing Weight
(aka: Leading)

Becoming a great leader is like losing weight; it's not complicated, it's just hard. Hear me out on this last analogy as we wrap up this book on leadership. Simplicity is key.

There are no secrets to losing weight, and you don't need to hire a personal trainer and nutritionist to remove the unwanted weight. Here it is: eat healthy (you know what this means, you just don't want to do it), workout consistently (just do *something* physically strenuous for sixty minutes a day, five days a week, and *never* skip), get good sleep (seven to eight hours a night, no exceptions), and drink lots of water. That's it. Period. Stick with this plan, and you'll lose weight and feel super awesome, I guarantee it. So why is there so much money spent each year by Americans trying to buy the latest fad in workout equipment, nutrition pills, and sleeping aids? Because most people don't put in the hard work of just doing it right, and then stay obstinately consistent at building those habits into a lifestyle. Instead, they cut corners, make excuses, and want an easier way to miraculously work.

My first year being the Character Coach at the brand-new high school where we currently live in Colorado, I put together a visual representation of what it meant to have to put in the hard work every day over an entire football season. I'm a big believer in the power of visualization, and sometimes to help teach younger kids the first steps toward the deeper levels of "seeing" your success mentally before it happens in reality, I use tangible visual means.

So I went to the banks around town (aka: I made my kids run in while I waited in the car since they were more convincing) and exchanged enough cash for pennies to add up to every day of practice we had scheduled, multiplied by the number of football players…which came to about six thousand pennies. You'd be surprised how hard-pressed banks are to have that many pennies anymore. I then took an old Air Force combat helmet and painted this phrase on it, "A penny a day, sweats complacency away." Lastly, I took all the pennies and poured them into a big jar and wrote "Potential" on the outside of it.

The first day of practice that season, I explained that after every practice each player would come over to the jar and grab one penny, which represented their hard work for the day toward earning their future potential. They would then drop it in the Air Force helmet which I had flipped upside down on the ground to receive the pennies. I explained that it was going to take all season for the entire team to transfer enough pennies from the "potential jar," into the helmet to fill it. I further explained that this is how real progress and change is made, through daily hard work over a long period of time. I wanted them to visually see their progress throughout the season as the jar slowly emptied and the helmet slowly filled up.

After the first couple weeks of practice, the helmet barely looked like it had anything in it. A penny per athlete isn't a lot. It definitely made the players realize the long haul that this was going to be. They kept at it, and about halfway through the season the helmet started gaining some weight and you could see it starting to fill up. By the end of the season, it was filled to the brim, and they had nearly emptied the entire "potential jar." You could also see the change that had occurred in these boys over an entire season of hard work.

The difference between being really good and being excellent is a super thin line. I love watching the Olympics. I have always been intrigued and amazed at how these humans have been able to push their minds to do what I consider superhuman physical feats. It's always so remarkable to me how the difference between first and fourth place is often razor thin. The difference between standing on the podium watching your country's flag be raised while an Olympic medal hangs around your neck and having to watch the medal ceremony from the crowd is almost discouraging. Yet, that fourth-place athlete is still in the top fraction of a percentile of world athletes, they're just not considered "superhuman" like those standing on the podium.

What continues to be something I research and try to refine the answer to, is how someone gains that razor thin edge over the competition. Even more baffling is how someone like Usain Bolt or Michael Phelps can *consistently* maintain that razor thin edge over their competition, year after year, in multiple events. It's a lifetime achievement for anyone, in any sport, to come home with a single Olympic gold medal. However, these two in particular have proven that it was not a fluke or that it just happened to be their day to race better than the competition. They have proven *consistent dominance.*

I use this curiosity of mental tenacity over physical limitations I witness during the Olympic games to keep me hungry for how to cross the bridge from really good leader to becoming an excellent leader.

Which brings me all the way back to my days of high school cross country.

Our high school team dominated the sport of cross country. During my four years at that high school (three of which I ran…I didn't start running until my sophomore year), the girls cross country team won State four years in a row. My junior year, we (the boys team) went undefeated all season and won State alongside our girls team. My senior year, we also went undefeated, were repeat District Champions, but then lost State by eight points (which in cross country scoring is like losing by a buzzer beater in basketball). That's a story for another day, but the point is our teams dominated. When I graduated in 2003, our legendary head coach, Tracy Harris, transferred to the adjacent new school that had just opened, and he has since continued to dominate with numerous State Championships (he is still coaching there, twenty years later). He is an amazing coach who ran in college and professionally, but more importantly has changed kids' lives by teaching them how to push themselves mentally, through physical means.

Coach Harris has his own catchphrase which he has used for decades: "Remain Curious." It was always somewhere on our seasonal team shirts and was a phrase all of us runners came to immortalize. It confused other teams who would read it on our gear and wonder exactly what it meant, which made it even better. We all came to understand the phrase nearly daily at practice when Coach Harris would see one of us starting to struggle and encourage us to "Remain Curious." It meant to always be discovering new levels of yourself and your abilities through wonder and hard work. He would say all the time how our bodies are capable of much more than our minds allow, so we needed to mentally strive to dig into the core of our will to push harder and curiously explore the next level within. It worked for me, and continues to be a phrase I use today, in nearly every facet of my life.

Coach Harris also taught me how to apply visualization to create real success. The day before every meet he would have the team slow jog the 3.1-mile (5K) course we were to race the next day. He had us point out places where the course could snag our foot with a tripping hazard, or where the hairpin turns were located, or where sand and hills could give you the advantage over the others not trained on how to leverage those differences in terrain to their advantage. He would then tell us to go home that night and as we were lying in bed, close our eyes and picture the course in our minds. He trained us to see those different aspects of the course and also visualize ourselves successfully dominating the course as we sprinted across the finish line. We essentially won every race before we even stepped up to the starting line on race day.

Another thing I remember Coach Harris saying on occasion was that other teams' coaches were constantly trying to figure out what our team did that gave us the consistent edge over the competition, year after year. He would say the same thing every time, that there was no secret to our success; we just worked harder than everyone else.

His simple phrase to "Remain Curious" and his matter-of-fact conclusion as to why we were better than other teams resonated with me at a young age. I have applied these simple concepts to my life ever since and have found great success.

Leadership is no different. Good, Better. Excellent.

The principles I've observed from these elite athletes and my own experiences have allowed me to conclude that these common denominators of consistent hard work and continuous improvement can be applied for anyone wanting to lead at a higher level, above their current rank and position.

There is no secret to becoming an excellent leader. You now have the recipe. This book is your recipe. Now put in the hard work.

••———————————••

I know I write bluntly and matter-of-fact, but I humbly admit my chasm of faults and incompetencies. Servant leadership is at my core and sometimes the way I write comes off as arrogant and self-centered. Those in my squadron know me better than that, but to the reader who has never met me, I don't want you getting the wrong impression and think you can just go around being an unnecessary ass. That's not what I am promoting and that's not what has caused me success.

Nevertheless, being a "badass" is much different than being an "unnecessary ass." So, take charge of your life and your career! Don't let others push you around. Work tirelessly applying the principles within this book while maintaining a healthy balance with the other very important parts of your life outside of the workplace.

I want to end by giving you one more phrase to repeat in your mind when you're attempting to apply these tactics during your journey of leading fearlessly above your rank, and when you start struggling to find the motivation to keep going when it gets hard. Because it will be very hard, most of the time.

By the time my family and I had moved to our current home in Colorado, we had moved fifteen times in fifteen years of marriage and lived in eight different states. Many of those moves were repeats back to a state we had already lived in, and a few were cross-town moves. I'm happy to report that at the time of writing this book, only the number of years married has increased since then. I think the number of years married would've been the one to stay at fifteen had I tried to increase the number of moves any further. Anyone who has moved an entire family, even just across town, knows how hard it is. Let alone crisscrossing the country with a family of six and a dog, and constantly uprooting from friends and communities you had just started to really integrate with.

To get through each move, we started using a phrase that we now use in our family for anytime we are about to do something we don't necessarily want to do: "We can do hard things!" It immediately helps us feel united as a family, that even if the trial is individual, you've got the other five of us backing you up and cheering you toward conquering whatever it is. We also have phrases such as "Make do" and "It's all part of the adventure (IAPOTA)" (my second son coined that one), and "There is no 'mine'," …but those are for another book.

"We can do hard things" must be part of your mental go-to mottos. Leadership is hard. That's why, just like there are thousands of dieting books, there are just as many books on leadership. The difference is you read my book not to find the easy way out, but because you are constantly looking to improve yourself. Maybe you picked my book because of the phrase "Leading fearlessly above your rank" and you want to learn how to be a leader *before* you have the official title. This is superb! You will find that your influence as an unofficial leader in your workplace will strengthen those around you and will ultimately lead you to the position you're already emulating. Once in that position of authority, those previous experiences will allow you to make the necessary official changes that will benefit your people long-term. The best part is, you *were* and *are* a consistent leader amongst your coworkers, with or without the title.

Maybe you picked up this book because you are currently in an official leadership position and already hold the appropriate rank and/or authority, but you want to find ways to make your organization progress from good to better…or better to excellent. Hell yeah! And guess what? Although you currently hold the correct rank for the commensurate position, you too need to always be operating at a level above your current rank. Unless you are the CEO or POTUS, you have a boss (and even the POTUS has a boss: the American people). Go take your boss's job and do it better! Always be improving.

Maybe you just wanted to learn how to be a better employee while trying to work with poor leaders. This is amazing and most commendable. You, too, can have a hugely profound influence on your work center by emulating what your official leader isn't. Not only will your fellow employees reach higher potential because of your non-positional influence, but maybe your leader will start seeing what they should be doing and make the necessary changes to become the leader you need them to be.

Whatever the reason for starting this book, don't be a damn quitter and not do the hard things necessary to become an excellent leader for your people when it starts getting tough or you start to become afraid. Take courage! Be bold! Have some grit, dammit! They deserve someone who is constantly self-improving and finding ways to better serve. Stay motivated through the rough times, knowing it will be worth it once you come out on the other side having never quit, never cut corners, and always followed your gut and heart. You can do hard things!

Lastly, always be able to answer your own personal "why" in life. If you aren't passionate about what you are doing in life, in any aspect, then make the necessary changes to pursue and achieve that passion! As Americans we have *every* opportunity available to us, so we should never have to settle. Do not settle! Setting your bar high and pursuing your dreams only strengthens the country as a whole.

Work your ass off for whatever it is you want and love. Stay positive through the darkness, get off your ass when you start feeling sorry for yourself, and remember the opportunities are out there for you to take—you just have to work super hard for them. During your journey to the top, never step on others to get there. Lift them up along with you and help them achieve potential they never realized they had. This will give you more satisfaction than you could have ever imagined.

Go forth, lead, and conquer. And, of course, *always*...stay FIRED UP!

About the Family Behind the Author

The Hochhalter family is no stranger to the sacrifices of military family life. Between fifteen moves (more to come), two deployments, and cumulative years' worth of TDYs, this family has dealt with their share of missed life events, leaving friends behind, and being the new kid. Nevertheless, they have steadfastly served their country with pride, honor, and an unwavering support for the difficult career path those who volunteer to serve must take. This book and the experiences written herein, would not have been made possible without the devotion of the home front.

www.ingramcontent.com/pod-product-compliance
Lightning Source LLC
Chambersburg PA
CBHW051636050426

42443CB00025B/379